T0257736

Scientific Researches in Atomic Force Microscopy

Scientific Researches in Atomic Force Microscopy

Edited by **Kate Wright**

New York

Published by NY Research Press,
23 West, 55th Street, Suite 816,
New York, NY 10019, USA
www.nyresearchpress.com

Scientific Researches in Atomic Force Microscopy
Edited by Kate Wright

© 2015 NY Research Press

International Standard Book Number: 978-1-63238-409-6 (Hardback)

This book contains information obtained from authentic and highly regarded sources. Copyright for all individual chapters remain with the respective authors as indicated. A wide variety of references are listed. Permission and sources are indicated; for detailed attributions, please refer to the permissions page. Reasonable efforts have been made to publish reliable data and information, but the authors, editors and publisher cannot assume any responsibility for the validity of all materials or the consequences of their use.

The publisher's policy is to use permanent paper from mills that operate a sustainable forestry policy. Furthermore, the publisher ensures that the text paper and cover boards used have met acceptable environmental accreditation standards.

Trademark Notice: Registered trademark of products or corporate names are used only for explanation and identification without intent to infringe.

Printed in the United States of America.

Contents

Permissions

List of Contributors

Preface

This book was inspired by the evolution of our times; to answer the curiosity of inquisitive minds. Many developments have occurred across the globe in the recent past which has transformed the progress in the field.

This book elucidates the scientific researches in the field of atomic force microscopy. The invention of the atomic force microscope (AFM) brought about drastic changes in the field of surface analysis. It proved to be a critical investigative resource and method, used for qualitative and quantitative study of surfaces with sub-nanometer resolutions. Additionally, samples analyzed through this microscope don't need prior preparation procedures. This prevents any alterations or adverse effects which may damage the sample while allowing a three dimensional study of the exterior surface. This book presents latest work by masters of this method across the globe. This method has found ready acceptance in procuring important information in a diverse spectrum of fields. Since its inception in 1986, it has found multiple uses across manufacturing, research and advancement fields.

The book was developed from a mere concept to drafts to chapters and finally compiled together as a complete text to benefit the readers across all nations. To ensure the quality of the content we instilled two significant steps in our procedure. The first was to appoint an editorial team that would verify the data and statistics provided in the book and also select the most appropriate and valuable contributions from the plentiful contributions we received from authors worldwide. The next step was to appoint an expert of the topic as the Editor-in-Chief, who would head the project and finally make the necessary amendments and modifications to make the text reader-friendly. I was then commissioned to examine all the material to present the topics in the most comprehensible and productive format.

I would like to take this opportunity to thank all the contributing authors who were supportive enough to contribute their time and knowledge to this project. I also wish to convey my regards to my family who have been extremely supportive during the entire project.

Editor

Atomic Force Microscopy in Optical Imaging and Characterization

Martin Veis[1] and Roman Antos[2]

[1]*Institute of Physics, Faculty of Mathematics and Physics, Charles University*
Institute of Biophysics and Informatics, 1st Faculty of Medicine, Charles University
[2]*Institute of Physics, Faculty of Mathematics and Physics, Charles University*
Czech Republic

1. Introduction

Atomic force microscopy (AFM) is a state of the art imaging system that uses a sharp probe to scan backwards and forwards over the surface of an object. The probe tip can have atomic dimensions, meaning that AFM can image the surface of an object at near atomic resolution. Two big advantages of AFM compared to other methods (for example scanning tunneling microscopy) are: the samples in AFM measurements do not need to be conducting because the AFM tip responds to interatomic forces, a cumulative effect of all electrons instead of tunneling current, and AFM can operate at much higher distance from the surface (5-15 nm), preventing damage to sensitive surfaces.

An exciting and promising area of growth for AFM has been in its combination with optical microscopy. Although the new optical techniques developed in the past few years have begun to push traditional limits, the lateral and axial resolution of optical microscopes are typically limited by the optical elements in the microscope, as well as the Rayleigh diffraction limit of light. In order to investigate the properties of nanostructures, such as shape and size, their chemical composition, molecular structure, as well as their dynamic properties, microscopes with high spatial resolution as well as high spectral and temporal resolving power are required. Near-field optical microscopy has proved to be a very promising technique, which can be applied to a large variety of problems in physics, chemistry, and biology. Several methods have been presented to merge the optical information of near-field optical microscopy with the measured surface topography. It was shown by (Mertz et al. (1994)) that standard AFM probes can be used for near-field light imaging as an alternative to tapered optical fibers and photomultipliers. It is possible to use the microfabricated piezoresistive AFM cantilevers as miniaturized photosensitive elements and probes. This allows a high lateral resolution of AFM to be combined with near-field optical measurements in a very convenient way. However, to successfully employ AFM techniques into the near-field optical microscopy, several technical difficulties have to be overcome.

Artificial periodical nanostructures such as gratings or photonics crystals are promising candidates for new generation of devices in integrated optics. Precise characterization of their lateral profile is necessary to control the lithography processing. However, the limitation of AFM is that the needle has to be held by a mechanical arm or cantilever. This restricts the access to the sample and prevents the probing of deep channels or any surface that isn't predominantly horizontal. Therefore to overcome these limitations the combination of AFM and optical scatterometry which is a method of determining geometrical (and/or material) parameters of patterned periodic structures by comparing optical measurements with simulations, the least square method and a fitting procedure is used.

2. AFM probes in near-field optical microscopy

In this section we review two experimental approaches of the near-field microscopy that use AFM tips as probing tools. The unique geometrical properties of AFM tips along with the possibility to bring the tip apex close to the sample surface allow optical resolutions of such systems to few tens of nanometers. These resolutions are not reachable by conventional microscopic techniques. For readers who are interested in the complex near-field optical phenomena we kindly recommend the book of (Novotny & Hecht (2006)).

2.1 Scattering-type scanning near-field optical microscopy

Scanning near field optical microscopy (SNOM) is a powerful microscopic method with an optical resolution bellow the Rayleigh diffraction limit. The optical microscope can be setup as either an aperture or an apertureless microscope. An aperture SNOM (schematically shown in Fig. 1(a)) uses a metal coated dielectric probe, such as tapered optical fibre, with a submicrometric aperture of diameter d at the apex. For the proper function of such probe it is necessary that d is above the critical cutoff diameter $d_c = 0.6\lambda/n$, otherwise the light propagation becomes evanescent which results in drastic λ dependent loss (Jackson (1975)). This cutoff effect significantly limits the resolution which can be achieved. The maximal resolution is therefore limited by the minimal aperture $d \approx \lambda/10$. In the visible region the 50 nm resolution is practically achievable (Hecht (1997)). With the increasing wavelength of illumination light, however, the resolution is decreased. This leads to the maximum resolution of $1\mu m$ in mid-infrared region, which is non usable for microscopy of nanostructures.

To overcome the limitations of aperture SNOM, one can use a different source of near field instead of the small aperture. This source can be a small scatter, such as nanoscopic particle or sharp tip, illuminated by a laser beam. When illuminated, these nanostructures provide an enhancement of optical fields in the proximity of their surface. This is due to a dipole in the tip which is induced by the illumination beam. This dipole itself induces a mirror dipole in the sample when the tip is brought very closely to its surface. Owing to this near-field interaction, complete information about the sample's local optical properties is determined by the elastically scattered light (scattered by the effective dipole emerging from the combination of tip and sample dipoles) which can be detected in the far field using common detectors. This is a basis of the scattering-type scanning near-field optical microscope (s-SNOM). There are two observables of practical importance in the detected signal: The absolute scattering efficiency and the material contrast (the relative signal change when probing nanostructures made from different materials). The detection of scattered radiation was first demonstrated in the microwave region by (Fee et al. (1989)) (although the radiation was confined in waveguide) and later demonstrated at optical frequencies by using an AFM tip as a scatterer (Zenhausern et al. (1995)) The principle of s-SNOM is shown in Figure 1(b). Both the optical and mechanical resolutions are determined by the radius of curvature a at the tip's apex and the optical resolution is independent of the wavelength of the illumination beam. To theoretically solve the complex problem of the realistic scattering of the illuminating light by an elongated tip in

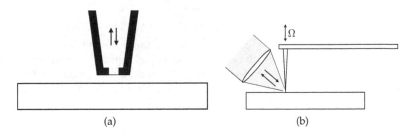

(a)   (b)

Fig. 1. Principles of aperture (a) and apertureless (b) scanning near-field optical microscopies.

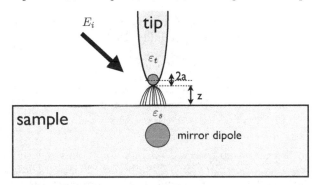

Fig. 2. Schematic view of the simplified theoretical geometry, where the tip was replaced by a small sphere at the tips apex. The sample response is characterized by an induced mirror dipole.

the proximity of the sample's surface it is necessary to use advanced electromagnetic theory, which is far beyond the scope of this chapter (readers are kindly referred to the work of (Porto et al. (2000))). However (Knoll & Keilmann (1999b)) demonstrated that the theoretical treatment based on simplified geometry can be used for quantitative calculation of the relative scattering when probing different materials. They have approximated the elongated probe tip by a polarizable sphere with dielectric constant ε_t, radius a ($a \ll \lambda$) and polarizability (Zayats & Richards (2009))

$$\alpha = 4\pi a^3 \frac{(\varepsilon_t - 1)}{(\varepsilon_t + 2)}. \tag{1}$$

This simplified geometry is schematically shown in Figure 2. The dipole is induced by an incident field E_i which is polarized parallel with the tip's axis (z direction). The incident polarization must have the z component. In this case the tip's shaft acts as an antenna resulting in an enhanced near-field (the influence of the incident polarization on the near-field enhancement was investigated by (Knoll & Keilmann (1999a))). This enhanced field exceeds the incident field E_i resulting in the indirect polarization of the sample with dielectric constant ε_s, which fills the half-space $z < 0$. Direct polarization of the sample by E_i is not assumed.

To obtain the polarization induced in the sample, the calculation is approximated by assuming the dipole as a point in the centre of the sphere. Then the near-field interaction between the tip dipole and the sample dipole in the electrostatic approximation can be described by the polarizability $\alpha\beta$ where

$$\beta = \frac{(\varepsilon_s - 1)}{(\varepsilon_s + 1)} \tag{2}$$

Note that the sample dipole is in the direction parallel to those in the tip and the dipole field is decreasing with the third power of the distance. Since the signal measured on the detector is created by the light scattered on the effective sample-tip dipole, it is convenient to describe the near-field interaction by the combined effective polarizability as was done by (Knoll & Keilmann (1999b)). This polarizability can be expressed as

$$\alpha_{eff} = \frac{\alpha(1 + \beta)}{1 - \frac{\alpha\beta}{16\pi(a+z)^3}}, \tag{3}$$

where z is the gap width between the tip and the sample. For a small particle, the scattered field amplitude is proportional to the polarizability (Keilmann & Hillenbrand (2004))

$$E_s \propto \alpha_{eff} E_i. \tag{4}$$

Since the quantities ε, β and α are complex, the effective polarizability can be generally characterized by a relative amplitude s and phase shift φ between the incident and the scattered light

$$\alpha_{eff} = s e^{i\varphi}. \tag{5}$$

The validity of the theoretical approach described above was determined by numerous s-SNOM studies published by (Hillenbrand & Keilmann (2002); Knoll & Keilmann (1999b); Ocelic & Hillenbrand (2004)). Good agreement between experimental and theoretical s-SNOM contrast was achieved.

Recalling the Equation (3) it is important to note that the change of the illumination wavelength will lead to changes in the scattering efficiency as the values of the dielectric constants ε_s and ε_t will follow dispersion relations of related materials. This allows to distinguish between different materials if the tip's response is flat in the spectral region of interest. Therefore the proper choice of the tip is important to enhance the material contrast and the resolution.

(Cvitkovic et al. (2007)) reformulated the coupled dipole problem and derived the formula for the scattered amplitude in slightly different form

$$E_s = (1 + r)^2 \frac{\alpha(1 + \beta)}{1 - \frac{\alpha\beta}{16\pi(a+z)^3}}. \tag{6}$$

They introduced Fresnel reflection coefficient of the flat sample surface. This is important to account for the extra illumination of the probe via reflection from the sample which was neglected in Equations (3) and (4).

The detected signal in s-SNOM is a mixture of the near-field scattering and the background scattering from the tip and the sample. Prior to the description of various experimental s-SNOM setups it is important to note how to eliminate the unwanted background scattering from the detector signal. For this purpose we have calculated the distance dependence of α_{eff}. The result is displayed in Figure 3. As one can see from this figure, the scattering is almost constant for distances larger than $2a$. On the other hand for very short distances (very closely to the sample) both the scattering amplitude and the scattering phase drastically increase. This

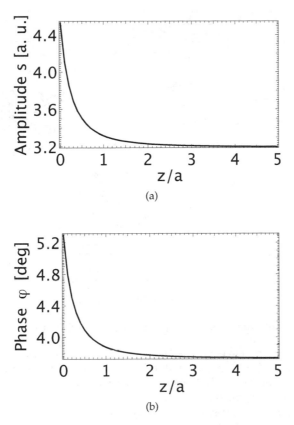

Fig. 3. Theoretically calculated dependence of the near-field scattering amplitude s (a) and phase φ (b) on the tip-sample distance z.

occurs for various materials with various dielectric constants demonstrating the near-field interaction.

When the tip is illuminated by a focused laser beam, only a small portion of the incident light reaches the gap between the tip and the sample and contributes to the near-field. Therefore the detected signal is mainly created by the background scattering. The nonlinear behavior of the $\alpha_{eff}(z)$ is employed to filter out the unwanted background scattering which dominates in the detected signal. This can be done if one employs tapping mode with a tapping frequency Ω into the experimental setup. The tapping of amplitude $\Delta z \approx a \approx 20nm$ modulates the near-field scattering much stronger than the background scattering. The nonlinear dependence of $\alpha_{eff}(z)$ will introduce higher harmonics in the detected signal. The full elimination of the background is done by demodulating the detector signal at the second or higher harmonic of Ω as was demonstrated by (Hillenbrand & Keilmann (2000)) and others.

There are various modifications of the s-SNOM experimental setup. Schematic views of interferometric s-SNOM experimental setups with heterodyne, homodyne and pseudohomodyne detection are displayed in Figure 4.

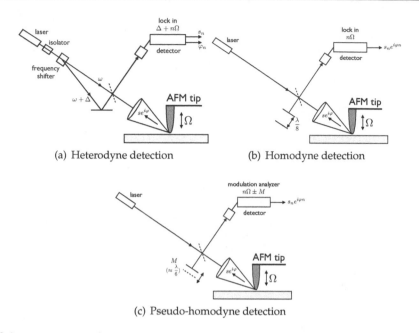

(a) Heterodyne detection   (b) Homodyne detection

(c) Pseudo-homodyne detection

Fig. 4. Schematic views of experimental interferometric s-SNOM setups

The heterodyne detection system developed by (Hillenbrand & Keilmann (2000)) uses a HeNe laser with output power of $\approx$ 1mW as the illumination source. The beam passes through the optical isolator to filter the back reflections from the frequency shifter. The frequency shifter creates a reference beam with the frequency shifted by $\Delta = 80$MHz which interferes with the backscattered light from the sample in a heterodyne interferometer. The detected intensity is therefore $I = I_{ref} + I_s + 2\sqrt{I_{ref}I_s}\cos(\Delta t + \varphi)$. The signal is processed in high frequency lock-in amplifier which operates on the sum frequency $\Delta + n\Omega$. Here n is the number of higher harmonic. The lock-in amplifier gives two output signals. One is proportional to scattering amplitude while the second is proportional to the phase of the detector modulation at frequency $\Delta + n\Omega$. When the order of harmonic n is sufficiently large the signal on the lock-in amplifier is proportional to s_n and φ_n. This means that using higher harmonics, one can measure pure near-field response directly. Moreover such experimental setup has optimized signal/noise ratio.

The influence of higher harmonic demodulation on the background filtering is demonstrated in Figure 5. In this figure the tip was used to investigate gold islands on Si substrate. For $n = 1$ the interference of different background contributions is clearly visible for $z > a$. Such interference may overlap with the important near-field interaction increase for $z < a$ which leads to a decrease of the contrast. Taking into account the second harmonic ($n = 2$) one can see a rapid decrease of the interference which allow near-field interactions to be more visible. For the third harmonic ($n = 3$) the near-field interaction becomes even steeper.

Because the tip is periodically touching the sample, a nonsinusiodal distortion of the taping motion can be created by the mechanical motion. This leads to artifacts in the final microscopic image which are caused by the fact that the higher harmonics $n\Omega$ are excited also by

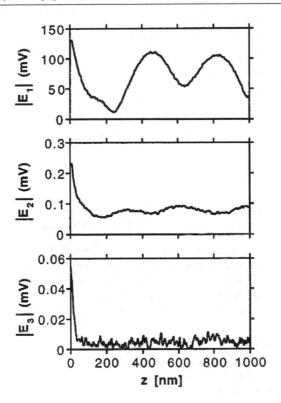

Fig. 5. Optical signal amplitude $|E_n|$ vs distance z between tip and Au sample, for different harmonic demodulation orders n. ©2002 American Institute of Physics (Hillenbrand & Keilmann (2002))

mechanical motion. These mechanical harmonics cause direct modulations of the optical signals resulting in a distorted image. (Hillenbrand et al. (2000)) demonstrated that the artifacts depend on the sample and the tapping characteristics (such as amplitude, etc.). They have found that these mechanical artifacts are negligible for small $\Delta z < 50nm$ and large setpoints $\Delta z / \Delta z_{free} > 0.9$.

In the mid-infrared region the appropriate illumination source was a CO_2 laser owing to its tunable properties from 9.2 to 11.2 μm. The attenuated laser beam of the power $\approx 10mW$ was focused by a Schwarzschild mirror objective (NA = 0.55) to the tip's apex. The polarization of the incident beam was, as in the previous case, optimized to have a large component in the direction of the tip shaft. This lead to a large enhancement of the near field interaction and increased the image contrast. The incident laser beam was split to create a reference which was reflected on a piezoelectrically controlled moveable mirror (Figure 4(b)). This mirror and the scattering tip created a Michelson interferometer. Using a homodyne detection the experimental setup was continuously switching the mirror between two positions. The first position corresponded to the maximum signal of the n-th harmonic at the lock-in amplifier (positive interference between the near-field scattered light and the reference beam) while the second position was moved by a $\lambda/8$ (90° shift of reference beam). With the experimental

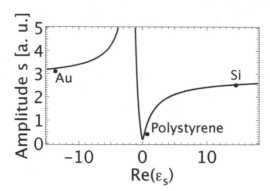

Fig. 6. Theoretically calculated the near-field scattering amplitude s as a function of the real part of ε_s. The imaginary part of ε_s was set to 0.1 and the tip was considered as Pt.

setup the detection of the amplitude and the phase of the near-field scattering was possible to detect, obtaining the near-field phase and amplitude contrast images. Further improvement of the background suppression was demonstrated by (Ocelic et al. (2006)) using a slightly modified homodyne detection with a sinusoidal phase modulation of the reference beam at frequency M (see Figure 4(c)). This lead to the complete reduction of the background interference.

As we already mentioned and as is clearly visible from the equations described above, the near-field scattering depends on the dielectric function of the tip and the sample. We have calculated the amplitude of the near-field scattering as a function of the real part of ε_s using Equation (3). The tip is assumed to be Pt ($\varepsilon_t = -5.2 + 16.7i$) and the sphere diameter $a = 20\text{nm}$. The result is depicted in Figure 6. The imaginary part of ε_s was set to 0.1. The inserted dots represent the data for different materials at illumination wavelength $\lambda = 633\text{nm}$ (Hillenbrand & Keilmann (2002)). As can be clearly seen from Figure 6, owing to different scattering amplitudes, a good contrast in the image of nanostructures consists of Au, Polystyrene and Si components should allow for easily observable images. Indeed, this was observed by (Hillenbrand & Keilmann (2002)) and is shown in Figure 7. The AFM topography image itself can not distinguish between different materials. However, due to the material contrast, it is possible to observe different material structures in the s-SNOM image. This is consistent with the theoretical calculation in Figure 6. The lateral resolution of the s-SNOM image in Figure 7 is 10 nm.

2.2 Tip enhanced fluorescence microscopy

Owing to its sensitivity to single molecules and biochemical compositions, fluorescence microscopy is a powerful method for studying biological systems. There are various experimental setups of fluorescent microscopes that exceed the Rayleigh diffraction criterion which limits the practical spatial resolution to $\approx 250\text{nm}$. Recent modifications of conventional confocal microscopy, such as $4 - \pi$ (Hell & Stelzer (1992)) or stimulated emission depletion (Klar et al. (2001)) microscopies have pushed the resolution to tens of nanometers. Although these techniques offer a major improvement in the field of fluorescent microscopy, they require high power laser beams, specially prepared fluorophores and provide slow performance (not suitable for biological dynamics).

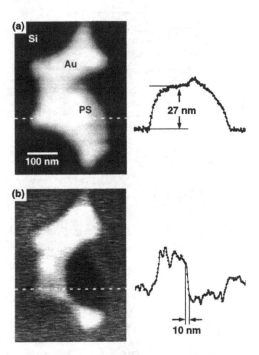

Fig. 7. Au island on Si observed in (a) topography, (b) optical amplitude $|E_3|$, with adjoining polystyrene particle. The line scans give evidence of purely optical contrast at 10 nm resolution, and of distinct near-field contrast levels for the three materials. ©2002 American Institute of Physics (Hillenbrand & Keilmann (2002))

Experimental setups of s-SNOM, as described in detail above, can be modified to sense fluorescence from nanoscale structures offering an alternative method to confocal microscopy. The near-field interaction between the sample and the tip causes the local increase of the one-photon fluorescence-excitation rate. The fluorescence is then detected by a single-photon sensitive avalanche photodiode. Such an experimental technique is called tip-enhanced fluorescence microscopy (TEFM) and its setup is schematically shown in Figure 8. There are two physical effects detected. The first one is an increase of detected fluorescence signal due to the near-field enhancement. The second one is the signal decrease due to the fluorescence quenching. These effects were demonstrated by various authors, for example by (Anger et al. (2006)). The fluorescence enhancement is proportional to the real part of the dielectric constant of the tip. On the other hand the fluorescence quenching is proportional to the imaginary part of the same dielectric function. Since these effects manifest themselves at short distances (bellow 20 nm), they can be used to obtain nanoscale resolution. Because the fluorescence enhancement leads to higher image contrast, silicon AFM tips are often used (due to their material parameters) for fluorescence studies of dense molecular systems.

In TEFM the illumination beam stimulates simultaneously a far-field fluorescence component S_{ff}, which is coming from direct excitation of fluorophores within the laser focus, and a near-field component S_{nf}, which is exited by a near-field enhancement. One can then define

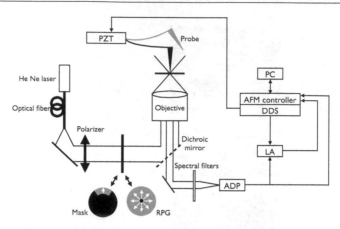

Fig. 8. Experimental setup of TEFM. RPG - radial polarization generator; PZT-piezoelectric transducer; ADP - avalanche photodiode; LA - lock-in amplifier; DDS - digital synthesizer.

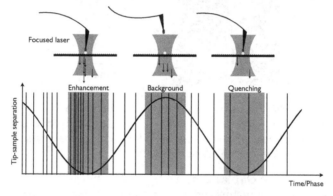

Fig. 9. Schematic picture of fluorescence modulation by AFM tip oscilation.

a contrast (C) of TEFM as

$$C = \frac{S_{nf}}{S_{ff}}. \tag{7}$$

Similarly to the s-SNOM setup it is possible to enhance the contrast and resolution of TEFM using a tapping mode in AFM and a demodulation algorithm for detected signal. Such process can be done by lock-in amplification. The scheme of the fluorescence modulation by an AFM tip oscillation is shown in Figure 9. When the tip is in the highest position above the sample no near-field interaction occurs. The detected signal is therefore coming from the background scattering excitation. If the tip is approaching the sample the fluorescence rate becomes maximally modified. The detected signal is either positive or negative depending on the fluorescence enhancement or quenching. TEFM example images of high density CdSe/ZnS quantum dots are shown in Figure 10. An improvement of the lateral resolution and contrast is clearly visible when using a TEFM with lock-in demodulation detection. The resolution of 10 nm, which is bellow the resolution of other fluorescence microscopies, demonstrates the main advantage of TEFM systems.

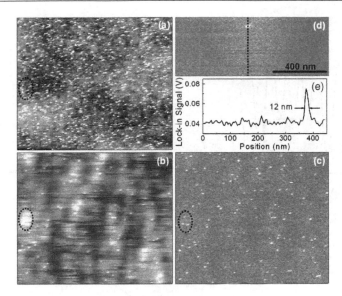

Fig. 10. High-resolution images of quantum dots. (a) AFM topography image; (b) photon-sum image; (c) TEFM image using lock-in demodulation. (a)–(c) are for a 5x5 μm^2 field of view. (d) TEFM image of a single quantum dot; (e) signal profile specified by the dotted line in (d). ©2006 American Institute of Physics (Xie et al. (2006))

3. AFM versus scatterometry

Recent advances of integrated circuits, including shortened dimensions, higher precision, and more complex shapes of geometric features patterned by modern lithographic methods, also requires higher precision of characterization techniques. This section briefly reviews some improvements of AFM and optical scatterometry, their comparison (with mutual advantages and disadvantages), and their possible cooperation in characterizing the quality of patterned nanostructures, especially in determining critical dimensions (CD), pattern shapes, linewidth roughness (LWR), or line edge roughness (LER).

3.1 AFM in the critical dimension (CD) mode with flared tips

It has been frequently demonstrated that accurate monitoring of sidewall features of patterned lines (or dots or holes) by AFM requires probe tips with special shapes and postprocessing algorithms to remove those shapes from the acquired images of the patterned profiles. A conventional AFM tip (with a conical, cylindrical, or intermediate shape) even with an infinitely small apex is only capable of detecting the surface roughness on horizontal surfaces (on the top of patterned elements or on the bottom of patterned grooves), but cannot precisely detect sidewall angles, LER, or particular fine sidewall features, as depicted in Fig. 11(a). Here the oscillation of the probe while scanning is only in the vertical direction so that the inverse profile of the tip is obtained instead of the correct sidewall shape. Only patterns with sidewall slopes smaller than the tip slopes (e.g., sinusoidal gratings) can be accurately detected after applying an appropriate image reconstruction transform such as the one shown by (Keller (1991)).

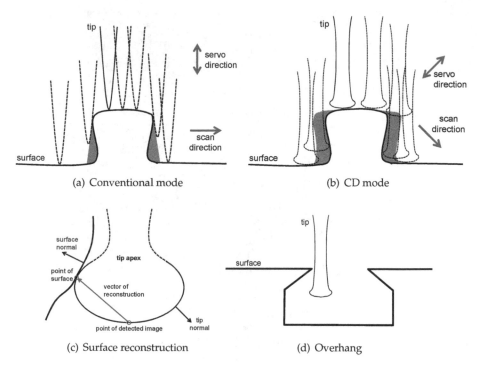

(a) Conventional mode   (b) CD mode

(c) Surface reconstruction   (d) Overhang

Fig. 11. AFM scanning of a patterned element in the conventional (a) and CD (b) modes with a conventional (a) and flared (b) tip apex. Postprocessing reconstruction of the surface uses the vector of reconstruction (c). The method can fail on overhang structures (d).

On the other hand, a CD tip, fabricated with a flared apex radius such as in (Liu et al. (2005)), can provide an accurate 3D patterned profile provided that it is applied in the CD mode. This mode, unlike the conventional deep trench mode where the tip only oscillates in the vertical direction, requires the tip to oscillate in both vertical and horizontal directions to follow the full surface topography for which multiple vertical points are possible for the same horizontal position, as depicted in Fig. 11(b). Analogously to conventional AFM scanning with a nonideal tip with a finite apex, the CD AFM scan also requires a postprocessing reconstruction of the real surface profile. An example of such a reconstruction, demonstrated by (Dahlen et al. (2005)), is the application of the reconstruction vector utilizing the fact that the normal to the surface is identical with the normal to the tip at each contact point, as displayed in Fig. 11(c). The method also utilizes algorithms of "reentrant" surface description.

The CD-AFM scanning can obviously fail for highly undercut surfaces for which the tip apex is not sufficiently flared, as visible in Fig. 11(d). However, such a structure can be advantageously used to carry out a topography measurement of the actual tip sidewall profile, as also described by (Dahlen et al. (2005)). Such a structure, designed solely for this reason, is called an overhang characterizing structure.

The advantages of the CD-AFM are that it is a nondestructive method (unlike cross-sectional SEM) which provides a direct image of the cross-sectional surface profile with relatively high precision. However, the image becomes truly direct after an appropriate postprocessing procedure removing the tip influence. Moreover, some profile features cannot be revealed such as the precise shape of the top sharp corners, the exact vertical positions and radii of fine sidewall features, and—most importantly—the real shape of sharp bottom corners of the grooves.

3.2 AFM used for line edge roughness (LER) characterization

As the dimensions of patterned structures shorten to the nanometer scale, the LER and the LWR become important characteristics. Following (Thiault et al. (2005)), we briefly define the LER and LWR as follows:

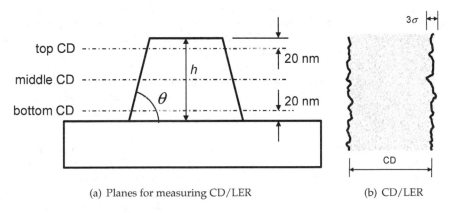

(a) Planes for measuring CD/LER   (b) CD/LER

Fig. 12. Definitions of measuring CD and LER in different planes for AFM in the CD mode.

The LWR is defined as three standard deviations (denoted 3σ) of the scanned linewidth variations at a height determined by the AFM user, while the CD represents the averaged value of the scanned dimension. Analogously, the LER is defined as its one-edge version, measured as three standard deviations of the variations from the straight line edge. Unfortunately, the CD-AFM scanning cannot accurately detect the linewidth at the very top and—especially—at the very bottom of the patterned grooves (which is due to the finite size of the tip apex). For this reason it is usually determined at some small height from the top (typically 20 nm, determined by the tip used), at the middle, and at some small height above the bottom (hence the top, middle, and bottom CD and LWR/LER, respectively). The corresponding planes are depicted in Fig. 12(a). The geometries for measuring the CD and the LER (3σ) at a chosen height are depicted in Fig. 12(b).

Although the LWR and LER at specified heights are also calculated from direct CD-AFM images, they can be affected by some further defects. As an example, consider a line whose both edges have equal LER values which are mutually uncorrelated. According to the statistical theory, $\sigma_{LWR}^2 = 2\sigma_{LER}^2$ should be valid, so that the LWR should be $2^{1/2}$ times higher than the LER. However, (Thiault et al. (2005)) have shown that the stage drift (breaking the relative position of the tip and sample) during long-time measurement affect the LER considerably more than the LWR, because the time between the detection of two adjacent edges is much shorter than the time between two sequential scans of the same edge.

3.3 Critical dimensions measured by scatterometry

Optical scatterometry, most often based on spectroscopic ellipsometry and sometimes combined with spectrophotometry (light intensity reflectance or transmittance), is an optical investigation method which combines optical measurements (typically in a wide spectral range, utilizing visible light with near ultraviolet and infrared edges) together with rigorous optical calculations. The spectra are calculated with varied input geometrical parameters of the patterned structure (optical CDs) and compared to the measured values to minimize the difference (optical error) as much as possible, employing the least square method for the optical error. The algorithm is usually referred to as the optical fitting procedure, and the obtained optical CDs are referred to as the optically fitted dimensions.

Various authors have presented the use of specular (0th-order diffracted) spectroscopic scatterometry to determine linewidths, periods, depths, and other fine profile features not accessible by AFM (such as the above mentioned bottom corners of grooves), as shown by (Huang & Terry Jr. (2004)). Obvious advantages of scatterometry is (besides non-destructiveness) higher sensitivity and no contact with any mechanical tool. Simply speaking, a photon examines the structure as it really stands and gives the true answer. Another advantage is the possibility to integrate an optical apparatus (most often a spectral ellipsometer) into a lithographer or deposition apparatus for the in situ monitoring of deposition and lithographic processes.

On the other hand, scatterometry has some disadvantages: Spectral measurements are indirect, and the measured spectra sometimes require very difficult analyses to reveal the real profile of the structure, which should be approximately known before starting the fitting procedure (to use it as an initial value). Moreover, the spectra can contain too many unknown parameters, or at least some vaguely known parameters. As an example, consider a grating made as periodic wires patterned from a Ta film deposited on a quartz substrate, which was optically investigated by (Antos et al. (2006)). The unknown parameters in the beginning were not only the geometrical dimensions and shape of the wires, but also the material properties of the Ta film, which were altogether correlated. Therefore, the first analysis was performed on a nonpatterned reference Ta film to determine the refractive index and extinction coefficient of Ta, as well as the thickness of the native oxide overlayer of Ta_2O_5. The obtained parameters were used as known constants within the second analysis, which was made on the Ta wire grating. This second analysis yielded the values of period, depth, linewidths, and the sidewall shape of the Ta wires. The sidewall shape was analytically approximated as paraboloidal, determined by two parameters: the top (smallest) linewidth (same as the bottom linewidth) and the middle (maximum) linewidth. Although the obtained geometry was revealed with higher precision than the geometry obtained by a direct method (unpublished SEM images in this case), all the obtained parameters were affected by a slight difference from the assumed paraboloidal sidewalls and by a native Ta_2O_5 overlayers that developed on the sidewalls (which were not taken into account in simulations). Each such difference or negligence from the real sample can contribute to discrepancy between the optical CDs and the real dimensions. Simply speaking, the advantage of high sensitivity can easily become a disadvantage, when the optical configuration is too sensitive to undesired features.

To illustrate the basic difference between conventional AFM and scatterometric measurements, consider a shallow rectangular grating patterned on the top of a 32-nm-thick

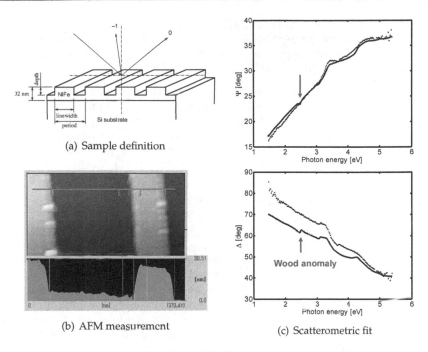

(a) Sample definition

(b) AFM measurement

(c) Scatterometric fit

Fig. 13. Investigation of a sample (a) by AFM (b) and scatterometry (c).

Permalloy (NiFe) film deposited on a Si substrate, with geometry depicted in Fig. 13(a) and with more details in (Antos et al. (2005d)). The comparison of nominal geometrical parameters (those intended by the grating manufacturer) with parameters determined by AFM [Fig. 13(b)] and scatterometry [based on spectroscopic ellipsometry performed at three angles of incidence, 60, 70, and 80°, the last of which is displayed in Fig. 13(c)] is listed in Table 1.

Parameter	nominal	AFM	scaled AFM	scatterometry
period	1000	1091.5	1000*	1000**
linewidth	500	359.4	329.3	307.2
NiFe thickness	32	—	—	—
relief depth	16	21.7	21.7	24.3

*fixed value
**just verified from the position of the Wood anomaly

Table 1. Comparison of grating geometrical parameters obtained by AFM and scatterometry, together with nominal ones (intended by the manufacturer).

Here the AFM measurement provides a direct image of the grating's relief profile (for our purposes to scan a shallow relief the conventional-mode AFM is appropriate), but the horizontal values (period and linewidth) are affected by a wrong-scale error (9 %, which is quite high). According to our experiences, the period of patterns made by lithographic processes is always achieved with high precision, so that we scale the horizontal AFM parameters to obtain the nominal 1000 nm period and to keep the same period-to-linewidth ratio (the verical depth is kept without change). The scaled AFM linewidth (329.3 nm, measured at the bottom) now corresponds well to the linewidth determined by scatterometry (307.2 nm); the 22 nm difference is probably due to the finite size of the apex of the AFM tip, which was previously explained in Fig. 11(a).

As visible in Fig. 13(c), the ellipsometric spectrum is not sensitive to the grating period, which is due to the shallow relief. For this reason the grating period could not be included in the fitting procedure. However, the period can be easily and with high precision verified by observing the spectral positions of Wood anomalies, which depend only on the period and the angle of incidence. In our case we observe the −1st-order Rayleigh wavelength (wavelength at which the −1st diffraction order becomes an evanescent wave) which provides precise verification of the nominal period of 1000 nm.

The 2.6 nm difference between the AFM and scatterometric values of the relief depth is difficult to explain. It might be again due the wrong vertical scale of the AFM measurement, or it might be due to some negligences of the used optical model such as the native top NiFe oxide overlayer or inaccurate NiFe optical parameters. Nevertheless, AFM and scatterometry differ from each other much less than how they differ from the nominal value, which indicates their adequacy. Finally, none of the methods were able to determine the full thickness of the deposited NiFe film (whose nominal value is 32 nm). While AFM was obviously disqualified in principle because it only detects the surface, optical scatterometry could provide this information if a reference sample (nonpatterned NiFe film simultaneously deposited on a transparent substrate) were investigated by energy transmittance measurement (again spectrally resolved for higher precision).

Another way to improve accuracy or the number of parameters to be resolved is to include higher diffraction orders in the analysis or to measure additional spectra such as magneto-optical spectroscopy (utilizing the magneto-optical anisotropy of NiFe). As an example, consider a grating made of Cr(2 nm)/NiFe(10 nm) periodic wires deposited on the top of a Si substrate. (Antos et al. (2005b)) used magneto-optical Kerr effect spectroscopy combining the analysis of the 0th and −1st diffraction order to determine the thicknesses of native oxide overlayers on the top of the Cr capping layer and on the top of Si substrate. It was shown that the 0th order of diffraction is more sensitive to features present both on wires and between them, whereas higher orders are more sensitive to differences between wires and grooves.

3.4 Line edge roughness determined by scatterometry

Besides measuring CDs, (Antos et al. (2005a)) have also shown that scatterometry is capable of evaluating the quality of patterning with respect to LER. Consider a pair of gratings similar to the one previously described, i.e., Cr(2 nm)/NiFe(10 nm) wires on the top of a Si substrate. AFM measurements of the two samples are shown in Fig. 14, where Sample 2 has obviously higher LER than Sample 1.

It is well known that p-polarized light is considerably more sensitive to surface features than s-polarized light. For this reason, magneto-optical Kerr effect spectroscopy in the −1st diffraction order was analyzed in a configuration where r_{pp} (amplitude reflectance for the p-polarization) is close to zero. Since the Kerr rotation and ellipticity are approximately equal to the real and imaginary components of the complex ratio r_{sp}/r_{pp} (here for the −1st order), they are very sensitive to the the wire edges. Fig. 15(a) displays experimental spectra

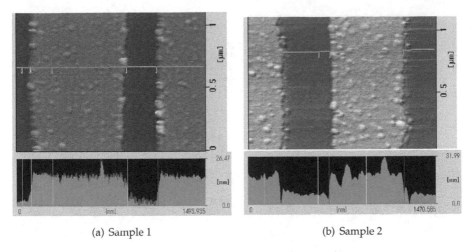

(a) Sample 1   (b) Sample 2

Fig. 14. AFM measurements of two Cr/NiFe wire samples with different LER.

measured on Sample 2 compared with two different models. First, the rigorous couple wave analysis (RCWA) is a rigorous method assuming diffraction on a perfect grating with ideal edges. Second, the local mode method (LMM) locally treats the grating structure as a uniform multilayer, neglecting thus the optical effect of the wire edges. Since none of the models corresponds well to the measured values, the reality is somewhere in the middle.

To include the effect of LER, we define a third model (optical LER method) as follows:

$$r_{pp}^{LER} = r_{pp}^{LMM} + \eta(r_{pp}^{RCWA} - r_{pp}^{LMM}) \tag{8}$$

where r_{pp}^{LMM} and r_{pp}^{RCWA} are reflectances calculated by the LMM and RCWA methods, and η is a parameter whose values can be between zero and one. For the case $\eta = 1$ the sample behaves as a perfect grating with ideal edges ($r_{pp}^{LER} = r_{pp}^{RCWA}$), so that LER is zero. For the opposite case $\eta = 0$ the grating behaves as a random formation of islands with the wire structure, with their relative area equal to the grating filling factor, so that LER is infinite or at least higher than the linewidth. In reality the η parameter will be somewhere in the middle and thus will provide the desired information about the LER. A fitting procedure carried on the two samples from Fig. 14 revealed the values of $\eta = 0.70$ for Sample 1 and $\eta = 0.53$ for Sample 2, which corresponds well to the obvious quality of the samples. The fitted spectra of the optical LER method are displayed in Fig. 15(b).

3.5 Joint AFM-scatterometry method

From the above comparisons, the mutual advantages and disadvantages of both AFM and scatterometry are obvious. For complex structures for which none of them can reveal all desired parameters when used solely, it is better to use them both. For instance, AFM can be used as the first method to determine the pattern shape and as many CDs as possible. Then, the obtained CDs are used as initial values for the optical fitting procedure in the frame of scatterometry, or some AFM parameters can be fixed (such as depth and top linewidth) and the remaining parameters can be fitted by scatterometry.

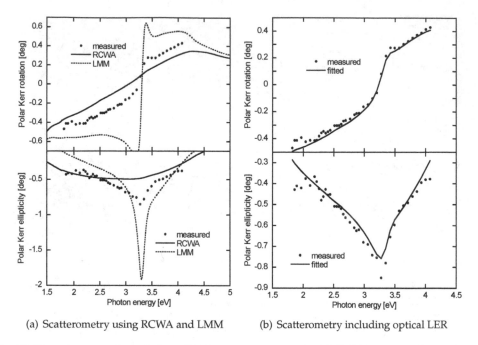

(a) Scatterometry using RCWA and LMM   (b) Scatterometry including optical LER

Fig. 15. Experiment and modeling of MO spectroscopy in the -1st diffraction order with p-polarized incident light assuming a perfect grating (a) and including optical LER (b).

As an example, consider a nearly-sinusoidal surface-relief grating patterned on the top of a thick epoxy layer with a refractive index close to the index of glass on which the epoxy was deposited. A sample of such a structure was investigated by (Antos et al. (2005c)) to obtain the following results: First, a detailed AFM scan of the surface provided the precise shape of the relief function, being something between a sinusoidal and a triangular function,

$$f(x) = a\frac{d}{2}\sin\frac{2\pi x}{\Lambda} + (1-a)\frac{2d}{\Lambda}x, \tag{9}$$

where a is a parameter of sharpness (the ratio between the sinusoidal and triangular shape), d is the depth of the grating, and Λ is its period. The AFM scan thus determined the period $\Lambda = 9365$ nm, depth of about $d = 700$ nm, and the parameter of sharpness $a = 0.6$. The period and the parameter of sharpness were then fixed as constants and used in a spectroscopic ellipsometry investigation to find more precisely the depth $d = 620$ nm and the values Δn of how much the epoxy's refractive index differs from the index of the glass substrate.

4. Conclusion

In this chapter we have shown that AFM tips can be used effectively as near-field probes in near-field microscopy. Using a proper experimental setup one can resolve nanostructures down to 10 nm indenpendently of the illumination wavelength, which can be chosen between visible and infrared region.

AFM and optical applications were also reviewed with respect to measuring geometries and dimensions of laterally patterned nanostructures. Both AFM in the CD mode and scatterometry were capable of providing valuable information on periodic relief profiles with different mutual advantages and disadvantages.

5. Acknowledgement

This work is part of the research plan MSM 0021620834 financed by the Ministry of Education of the Czech Republic and was supported by the Grant Agency of the Czech Republic (no. P204/10/P346 and 202/09/P355) and a Marie Curie International Reintegration Grant (no. 224944) within the 7th European Community Framework Programme.

6. References

Anger, P., Bahradwaj, P. & Novotny, L. (2006). Enhancement and quenching of single-molecule fluorescence, *Phys. Rev. Lett.* Vol. 96(No. 11): 113002.

Antos, R., Mistrik, J., Yamaguchi, T., Visnovsky, S., Demokritov, S. O. & Hillebrands, B. (2005a). Evaluation of the quality of Permalloy gratings by diffracted magneto-optical spectroscopy, *Opt. Express* Vol. 13(No. 12): 4651–4656.

Antos, R., Mistrik, J., Yamaguchi, T., Visnovsky, S., Demokritov, S. O. & Hillebrands, B. (2005b). Evidence of native oxides on the capping and substrate of permalloy gratings by magneto-optical spectroscopy in the zeroth- and first-diffraction orders, *Appl. Phys. Lett.* Vol. 86(No. 23): 231101.

Antos, R., Ohlidal, I., Franta, D., Klapetek, P., Mistrik, J., Yamaguchi, T. & Visnovsky, S. (2005). Spectroscopic ellipsometry on sinusoidal surface-relief gratings, *Appl. Surf. Sci.* Vol. 244(No. 1-4): 221–224.

Antos, R., Pistora, J., Mistrik, J., Yamaguchi, T., Yamaguchi, S., Horie, M., Visnovsky, S. & Otani, Y. (2006). Convergence properties of critical dimension measurements by spectroscopic ellipsometry on gratings made of various materials, *J. Appl. Phys.* Vol. 100(No. 5): 054906.

Antos, R., Veis, M., Liskova, E., Aoyama, M., Hamrle, J., Kimura, T., Gustafik, P., Horie, M., Mistrik, J., Yamaguchi, T., Visnovsky, S. & Okamoto, N. (2005). Optical metrology of patterned magnetic structures: deep versus shallow gratings, *Proc. SPIE* Vol. 5752(No. 1-3): 1050–1059.

Cvitkovic, A., Ocelic, N. & Hillenbrand, R. (2007). Analytical model for quantitative prediction of material contrasts in scattering-type near-field optical microscopy, *Opt. Express* Vol. 15(No. 14): 8550.

Dahlen, G., Osborn, M., Okulan, N., Foreman, W., Chand, A. & Foucher, J. (2005). Tip characterization and surface reconstruction of complex structures with critical dimension atomic force microscopy, *J. Vac. Sci. Technol. B* Vol. 23(No. 6): 2297–2303.

Fee, M., Chu, S. & Hänsch, T. W. (1989). Scanning electromagnetic transmission line microscope with sub-wavelength resolution, *Optics Communications* Vol. 69(No. 3-4): 219–224.

Hecht, B. (1997). Facts and artifacts in near-field microscopy, *J. Appl. Phys.* Vol. 81(No. 6): 2492–2498.

Hell, S. W. & Stelzer, E. H. K. (1992). Fundamental improvement of resolution with a 4Pi-confocal fluorescence microscope using 2-photon excitation, *Opt. Commun.* Vol. 93(No. 5-6): 277–282.

Hillenbrand, R. & Keilmann, F. (2000). Complex optical constants on a subwavelength scale, *Phys. Rev. Lett.* Vol. 85(No. 14): 3029–3032.

Hillenbrand, R. & Keilmann, F. (2002). Material-specific mapping of metal/semiconductor/dielectric nanosystems at 10 nm resolution by backscattering near-field optical microscopy, *Appl. Phys. Lett.* Vol. 80(No. 1): 25–27.

Hillenbrand, R., Stark, M. & Guckenberger, R. (2000). Higher-harmonics generation in tapping-mode atomic-force microscopy: Insights into the tip-sample interaction, *Appl. Phys. Rev.* Vol. 76(No. 23): 3478–3480.

Huang, H.-T. & Terry Jr., F. L. (2004). Spectroscopic ellipsometry and reflectometry from gratings (Scatterometry) for critical dimension measurement and in situ, real-time process monitoring, *Thin Solid Films* Vol. 455-456(No. 1-2): 828–836.

Jackson, J. D. (1975). *Classical Electrodynamics*, John Wiley, New York.

Keilmann, F. & Hillenbrand, R. (2004). Near-field microscopy by elastic light scattering from a tip, *Phil. Trans. R. Soc. Lond. A* Vol. 362(No. 1817): 787–805.

Keller, D. (1991). Reconstruction of STM and AFM images distorted by finite-size tips, *Surf. Sci.* Vol. 253(No. 1-3): 353–364.

Klar, T. A., Engel, E. & Hell, S. W. (2001). Breaking Abbe's diffraction resolution limit in fluorescence microscopy with stimulated emission depletion beams of various shapes, *Phys. Rev. E* Vol. 64(No. 6): 066613.

Knoll, B. & Keilmann, F. (1999a). Mid-infrared scanning near-field optical microscope resolves 30 nm, *Journal of Microscopy* Vol. 194(No. 2-3): 512–515.

Knoll, B. & Keilmann, F. (1999b). Near-field probing of vibrational absorption for chemical microscopy, *Nature* Vol. 399(No. 6732): 134–137.

Liu, H., Klonowski, M., Kneeburg, D., Dahlen, G., Osborn, M. & Bao, T. (2005). Advanced atomic force microscopy probes: Wear resistant designs, *J. Vac. Sci. Technol. B* Vol. 23(No. 6): 3090–3093.

Mertz, J., Hipp, M., Mlynek, J. & Marti, O. (1994). Facts and artifacts in near-field microscopy, *Appl. Phys. Lett.* Vol. 81(No. 18): 2338–2340.

Novotny, L. & Hecht, B. (2006). *Principles of nano-optics*, Cambridge Univeristy Press, Cambridge.

Ocelic, N. & Hillenbrand, R. (2004). Subwavelength-scale tailoring of surface phonon polaritons by focused ion-beam implantation, *Nature Materials* Vol. 3(No. 9): 606–609.

Ocelic, N., Huber, A. & Hillenbrand, R. (2006). Pseudoheterodyne detection for background-free near-field spectroscopy, *Appl. Phys. Lett* Vol. 89(No. 10): 101124.

Porto, J. A., Carminati, R. & Greffet, J. J. (2000). Theory of electromagnetic field imaging and spectroscopy in scanning near-field optical microscopy, *J. Appl. Phys.* Vol. 88(No. 8): 4845–4850.

Thiault, J., Foucher, J., Tortai, J. H., Joubert, O., Landis, S., & Pauliac, S. (2005). Line edge roughness characterization with a three-dimensional atomic force microscope: Transfer during gate patterning processes, *J. Vac. Sci. Technol. B* Vol. 23(No. 6): 3075–3079.

Xie, C., Mu, C., Cox, J. R. & Gerton, J. M. (2006). Tip-enhanced fluorescence microscopy of high-density samples, *Appl. Phys. Lett.* Vol. 89: 143117.

Zayats, A. & Richards, D. (2009). *Nano-optics and near-field optical microscopy*, Artech House, Norwood.

Zenhausern, F., Martin, Y. & Wickramasinghe, H. K. (1995). Scanning interferometric apertureless microscopy: Optical imaging at 10 Angstrom resolution, *Science* Vol. 269(No. 5227): 1083–1085.

Magnetic Force Microscopy: Basic Principles and Applications

F.A. Ferri[1], M.A. Pereira-da-Silva[1,2] and E. Marega Jr.[1]
[1]Instituto de Física de São Carlos, Universidade de São Paulo, São Carlos
[2]Centro Universitário Central Paulista, UNICEP, São Carlos
Brazil

1. Introduction

Magnetic force microscopy (MFM) is a special mode of operation of the atomic force microscope (AFM). The technique employs a magnetic probe, which is brought close to a sample and interacts with the magnetic stray fields near the surface. The strength of the local magnetostatic interaction determines the vertical motion of the tip as it scans across the sample.

MFM was introduced shortly after the invention of the AFM (Martin & Wickramasinghe, 1987), and became popular as a technique that offers high imaging resolution without the need for special sample preparation or environmental conditions. Since the early 1990s, it has been widely used in the fundamental research of magnetic materials, as well as the development of magnetic recording components. MFM detects the quantity that is of particular interest for the magnetic recording process, namely the magnetic stray field produced by a magnetized medium or by a write head. The magnetic transition geometry and stray field configuration in longitudinal recording media is illustrated in Fig. 1 (Rugar et al., 1990). Nowadays, the main developments in MFM are focused on the quantitative analysis of data, improvement of resolution, and the application of external fields during measurements (Schwarz & Wiesendanger, 2008).

The interpretation of images acquired by MFM requires knowledge about the specific near-field magnetostatic interaction between probe and sample. Therefore, this subject will be briefly discussed hereafter. Other topics to be considered are the properties of suitable probes, the achievable spatial resolution, and the inherent restrictions of the method. More detailed information can be found, e.g., in articles by Rugar et al., Porthun et al. and Hartmann. Valuable information can also be found in the works of Koch and Hendrych et al.

In the present chapter, we will also demonstrate some applications of the technique made by our research group in the study of magnetic vortices formation in sub-microsized structures, as well as further magnetic properties, of Si and Ge-based magnetic semiconductors thin films.

2. Basics of magnetic contrast formation

The operating principle of MFM is the same as in AFM. Both static and dynamic detection modes can be applied, but mainly the dynamic mode is considered here because it offers

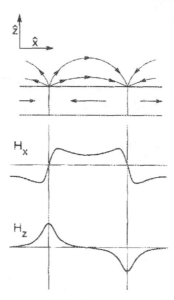

Fig. 1. Geometry of the magnetic stray field above a longitudinal magnetic medium (upper). Typical variation of the H_x and H_z components above the medium (lower) (Rugar et al., 1990)

better sensitivity. The cantilever (incorporating the tip) is excited to vibrate close to its resonance frequency, with a certain amplitude and a phase shift with respect to the drive signal. The deflection sensor of the microscope monitors the motion of the tip. Under the influence of a probe-sample interaction, the cantilever behaves as if it had a modified spring constant, $c_F = c - \partial F / \partial z$, where c is the natural spring constant and $\partial F / \partial z$ is the derivative of the interaction force relative to the perpendicular coordinate z. It is assumed that the cantilever is oriented parallel to the sample surface.

An attractive interaction with $\partial F / \partial z > 0$ will effectively make the cantilever spring softer, so that its resonance frequency will decrease. A shift in resonance frequency will lead to a change of the oscillation amplitude of the probe and of its phase. All of these are measurable quantities that can be used to map the lateral variation of $\partial F / \partial z$. The most common detection method uses the amplitude signal and is referred to as amplitude modulation (AM). The cantilever is driven slightly away from resonance, where the slope of the amplitude-versus-frequency curve is high, in order to maximize the signal obtained from a given force derivative.

Measurement sensitivity, or the minimum detectable force derivative, has an inverse dependence on the Q value of the oscillating system (Hartmann, 1999). Therefore, a high Q value might seem advantageous, but this has the drawback that it increases the response time of the detection system. In situations where Q is necessarily high, for example when scanning in vacuum, a suitable alternative is the frequency modulation (FM) technique (Porthun et al., 1998; Hartmann, 1999). In this method the cantilever oscillates directly at its resonance frequency by using a feedback amplifier with amplitude control.

The force derivative $\partial F / \partial z$ can originate from a wide range of sources, including electrostatic probe-sample interactions, van der Waals forces, damping, or capillary forces (Porthun et al., 1998). However, MFM relies on those forces that arise from a long-range magnetostatic coupling between probe and sample. This coupling depends on the internal magnetic structure of the probe, which greatly complicates the mechanism of contrast formation.

In general, a magnetized body, brought into the stray field of a sample, will have the magnetic potential energy E (Porthun et al., 1998):

$$E = -\mu_0 \int \vec{M}_{tip} \cdot \vec{H}_{sample} dV_{tip} \tag{1}$$

where μ_0 is the vacuum permeability. The force acting on an MFM tip can thus be calculated by:

$$\vec{F} = -\vec{\nabla}E = \mu_0 \int \vec{\nabla}\left(\vec{M}_{tip} \cdot \vec{H}_{sample}\right) dV_{tip} \tag{2}$$

The integration has to be carried out over the tip volume, or rather its magnetized part as illustrated in Fig. 2. Simplified models for the tip geometry and its magnetic structure are often used in order to make such calculations feasible. Another equivalent approach is to start the simulation with the tip stray field and to integrate over the sample volume (Porthun et al., 1998). According to Newton's third law, the force acting on the sample in the field of the tip is equal in magnitude to $\vec{F}$ in the previous equation:

$$\vec{F} = \mu_0 \int \vec{\nabla}\left(\vec{M}_{sample} \cdot \vec{H}_{tip}\right) dV_{sample} \tag{3}$$

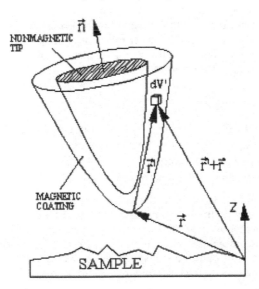

Fig. 2. Modelled MFM tip having a magnetic coating on a non-magnetic core. Parameters for integration are indicated (Koch, 2005)

The magnetostatic potential $\phi_s(\vec{r})$ created by any ferromagnetic sample can be calculated from its magnetization vector field $\vec{M}_s(\vec{r}')$ (Hartmann, 1999):

$$\phi_s(\vec{r}) = \frac{1}{4\pi}\left[\int\frac{d^2\vec{s}'\cdot\vec{M}_s(\vec{r}')}{|\vec{r}-\vec{r}'|} - \int d^3\vec{r}'\,\frac{\vec{\nabla}\cdot\vec{M}_s(\vec{r}')}{|\vec{r}-\vec{r}'|}\right] \qquad (4)$$

where $\vec{s}'$ is an outward normal vector from the sample surface. The first (two-dimensional) integral covers all surface charges created by magnetization components perpendicular to the surface, while the second (three-dimensional) integral contains the volume magnetic charges resulting from interior divergences of the magnetization vector field. The sample stray field is then given by $\vec{H}_{sample}(\vec{r}) = -\vec{\nabla}\phi_s(\vec{r})$, which can be substituted in Equation (2) to calculate the interaction force $\vec{F}$. In static mode the instrument detects the vertical component of the cantilever deflection, $F_d = \vec{n}\cdot\vec{F}$, where $\vec{n}$ is an outward unit normal from the cantilever surface. In the dynamic mode the compliance component or force derivative $F_{d'}(\vec{r}) = (\vec{n}\cdot\vec{\nabla})[\vec{n}\cdot\vec{F}(\vec{r})]$ is detected (Hartmann, 1999).

A limitation in the use of MFM is that the magnetic configuration of the sensing probe is rarely known in detail. Although the general theory of contrast formation still holds, it is not possible to model the measured signal from first principles for an unknown domain structure of the magnetic probe. As a consequence, MFM can generally not be performed in a quantitative way, in the sense that a stray field would be detected in absolute units. Furthermore, because MFM is sensitive to the strength and polarity of near-surface stray fields produced by ferromagnetic samples, rather than to the magnetization itself, it is usually not straightforward to deduce the overall domain topology from an MFM image. The problem of reconstructing a concrete arrangement of inner and surface magnetic charges from the stray fields they produce is not solvable. MFM can, however, be used to compare the experimentally detected stray field variation of a micromagnetic object to that obtained from certain model calculations. This often enables to at least classify the magnetic object under investigation (Hartmann, 1999). Thus, even without detailed quantitative analysis, the qualitative information collected by the microscope can be very useful (Rugar et al., 1990).

3. Modelling the MFM response

If one wants to analyze the force derivative $F_{d'}(\vec{r})$ using Equations (2) and (4), then a model of the tip shape and magnetization must be constructed. Various levels of complexity are possible. Most models assume that both the tip and the sample are ideally hard magnetic materials, with a magnetization that is unaffected by the stray field from the other.

The simplest way to model a tip is with the point-probe approximation (Hartmann, 1999). The effective monopole and dipole moments of the probe are projected into a fictitious probe of infinitesimal size that is located a certain distance away from the sample surface. The unknown magnetic moments as well as the effective probe-sample separation are treated as free parameters to be fitted to experimental data. The force acting on the probe, which is immersed in the near-surface sample microfield, is given by (Hartmann, 1999):

$$\vec{F} = \mu_0\left(q + \vec{m}\cdot\vec{\nabla}\right)\vec{H} \qquad (5)$$

where q and $\bar{m}$ are the effective monopole and dipole moments of the probe.

The point-probe approximation yields satisfactory results in many cases of MFM contrast interpretation. However, a far more realistic approach can be achieved by considering the extended geometry of a probe. An example is the pseudodomain model (Hartmann, 1999), in which the unknown magnetization vector field near the probe apex, with its entire surface and volume charges, is modelled by a homogeneously magnetized prolate spheroid of suitable dimensions. The magnetic response of the probe outside this imaginary domain is neglected. This pseudodomain model allows interpretation of most results obtained by MFM on the basis of bulk probes. For probes with a different geometry, for example those where the magnetic region is confined to a thin layer, other appropriate models have been developed (Rasa et al., 2002).

Fig. 3 shows both the measured and calculated MFM response across a series of 5 μm longitudinal bits (Rugar et al., 1990). The signal was recorded as a constant force derivative contour. In this particular case, the tip was modelled as a uniformly magnetized truncated cone with a spherical cap, in agreement with the shape as observed by electron microscopy (Rugar et al., 1990). Note that for in-plane magnetized samples, interdomain boundaries are the only sources of magnetic stray field that can be externally detected by MFM. On the other hand, samples with perpendicular magnetic anisotropy produce extended surface charges that correspond to the upward and downward pointing domain magnetization. In this case the near-surface stray field is directly related to the domain topology (Hartmann, 1999).

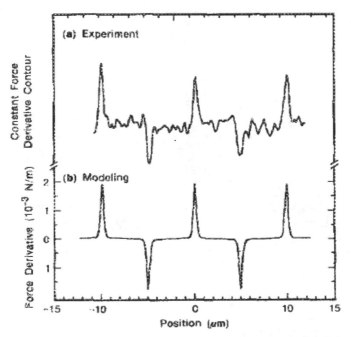

Fig. 3. (a) Contour of constant force derivative measured on a 5 μm bit sample. (b) Corresponding model calculation of magnetic force derivative (adapted from Rugar et al., 1990)

Usually, the MFM response of a certain tip-sample configuration is calculated by an integration in the spatial domain, e.g., over the sample volume. Porthun et al. have proposed a different formalism, where the problem is approached in the frequency domain. This has the advantage that it shows some characteristics of the imaging process more clearly. To be specific, the sample magnetization distribution is split up into harmonics, each having a spatial wavelength λ and wavenumber $k = 2\pi / \lambda$. The wavelength measures the length scale over which the magnetization vector goes through a complete rotation. Frequency components of the magnetic potential and the stray field are calculated separately. Then, the magnetic signal can be determined using Equation (2) for each of the stray field harmonics. For a specific (and simplified) tip-sample geometry (Fig. 4), the detected MFM signal is obtained by summing over all frequency components of the force derivative. The resulting signal, expressed in terms of sample magnetization and spatial frequency, forms a tip transfer function for the imaging process. An important observation is that the transfer function shows an exponential decay, $\exp(-kz_0)$, with increasing tip-sample distance z_0. It is thus crucial for high resolution to keep the tip-sample distance as small as possible. In addition, the dimensions (length, width, thickness) of a bar-type tip lead to specific decay rates both at high and low spatial frequencies. The latter illustrates that the finite size of a tip plays an important role in the imaging process. Therefore, a simple point-probe approximation is not sufficient to clarify how high and low spatial frequencies are attenuated. In the context of such a frequency domain description, the resolution can be defined as a minimum detectable wavelength which is determined by the noise limit of the detector system.

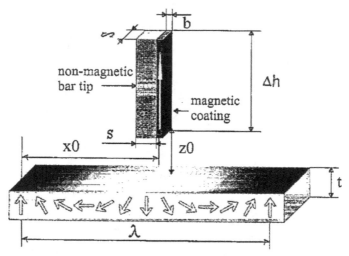

Fig. 4. One-dimensional model for the MFM measurement process (Porthun et al., 1998)

4. Requirements for MFM tips

The cantilever/tip assembly is obviously the critical element of a magnetic force microscope. Unlike in scanning tunnelling microscopy (STM) and repulsive-mode AFM, the tip shape is important due to the long-range nature of magnetic forces (Rugar et al., 1990). Originally, electrochemically etched wires of cobalt or nickel were used as cantilevers (Martin & Wickramasinghe, 1987). Thanks to the widespread use of AFM, cantilevers with integrated sharp tips are now fabricated in large numbers out of silicon-based materials. These tips can be coated with a thin layer of magnetic material for the purpose of MFM observations. A lot of effort has been spent on the optimization of magnetic tips in order to get quantitative information from MFM data (Rugar et al., 1990; Porthun et al., 1998; Hartmann, 1999). The problem is that in the coating of conventional tips, a pattern of magnetic domains will arrange, which reduces the effective magnetic moment of the tip. The exact domain structure is unknown and can even change during MFM operation. Nevertheless, some information on the magnetization state of selected probes has been acquired using electron holography (Rugar et al., 1990; Hartmann, 1999).

The spatial resolution in MFM imaging is related to the tip-sample distance, but also to the magnetized part of the tip that is actually exposed to the sample stray field. Thus in order to improve lateral resolution, it is beneficial to restrict the magnetically sensitive region to the smallest possible size. Ideally the effective volume of the probe would consist of a small single-domain ferromagnetic particle located at the probe apex. So-called supertips have been developed based on this idea (Hartmann, 1999). However, there is a physical lower limit for the dimensions because an ultra-small particle becomes superparamagnetic.

The demand for a strong signal, produced by a small sensitive volume, indicates the need to maximize the magnetic moment in the tip. For this reason a single domain tip will give the best results and is also easier to describe theoretically. Materials with a high saturation magnetization should be used in order to limit the required volume. The well-defined magnetic state of a tip should be stable during scanning, and it should interfere as little as possible with the sample magnetization. A high switching field of the tip can be realized through the influence of shape anisotropy (Porthun et al., 1998; Hartmann, 1999), which forces the magnetization vector field near the probe apex to align with its axis of symmetry. Eventually, the smallest detail from which a sufficient signal-to-noise ratio can be gained is determined by the sensitivity of the deflection sensor, as well as the noise characteristics of the cantilever (Porthun et al., 1998).

In the present work, we have employed etched silicon tips of the MESP type supplied by Bruker. These are standard probes for MFM, and have a pyramidal geometry (Fig. 5). The magnetic coating consists of ~ 10–150 nm of Co/Cr alloy (exact thickness and composition of the coatings are undisclosed). The cantilever has a length L of approximately 225 μm. As a result, the resonance frequency f_0 is about 75 kHz. The coating has a coercivity of ~ 400 Oe and a magnetic moment of 1×10^{-13} emu. In order to ensure a predominant orientation of the magnetic vector field along the major probe axis, the thin film probes were magnetized (along the cantilever) prior to taking measurements. The Digital Instruments company offers a magnetizing device that possesses a permanent magnet. This apparatus ensures that the distance from the magnet to the tip is always the same in different magnetization procedures. Thus, taking into account that the magnetic field lines are dependent on the distance, the reproducibility is then guaranteed.

Fig. 5. Scanning electron images of AFM probes like the ones used for MFM. The probes are coated with a magnetic thin film. Specifications are mentioned in the text (Bruker Corporation, 2011)

5. Imaging procedure

As in AFM scanning, the detector signal can be fed back to the scanner z actuator. This mode of operation is called constant signal mode, in contrast to the open-loop or constant distance mode. The constant signal mode is robust and allows an accurate tracking of the sample surface, but it also presents a few problems. For example, the magnetic signal can be positive or negative, while stable feedback is only possible when the interaction does not change sign. This makes it necessary to bias the signal: the application of a voltage between the sample and the tip introduces an additional (electrostatic) force. Another problem of this mode is that the magnetic and non-magnetic interactions are mixed. The mixing ratio depends on the tip-sample distance which itself depends on the magnetic interaction. This makes the contributions very difficult to separate. For operation in air, it is known that the interaction with the surface contaminant layer and the damping (in dynamic mode) have a stronger influence on the tip than the van der Waals interaction (Porthun et al., 1998).

Quantitative data about the sample stray field can only be derived from MFM images when topographic signal contributions are not included. This is especially important when the tip is brought very close to the sample (in order to improve resolution), since non-magnetic forces become increasingly stronger. The solution to this problem is to keep the topography influence constant by letting the tip follow the surface height profile (Porthun et al., 1998). This constant distance mode places higher demands on instrument stability, because it is sensitive to drift. In the Digital Instruments microscope (Nanoscope 3A Multimode), the specific method employed to separate signal contributions is called lift mode (Fig. 6). It involves measuring the topography on each scan line in a first scan (left panel), and the magnetic information in a second scan of the same line (right panel). The difference in height Δh between the two scans, the so-called lift height, is selected by the user. Topography is measured in dynamic AM mode and the data is recorded to one image. This height data is also used to move the tip at a constant local distance above the surface during the second (magnetic) scan line, during which the feedback is turned off. In theory, topographic contributions should be eliminated in the second image.

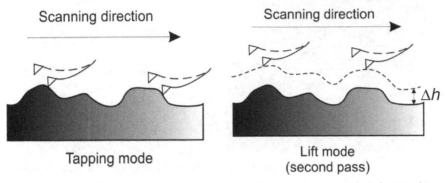

Fig. 6. Outline of the lift mode principle. Magnetic information is recorded during the second pass (right panel). The constant height difference between the two scan lines is the lift height Δh (adapted from Hendrych et al., 2007)

Magnetic data can be recorded either as variations in amplitude, frequency, or phase of the cantilever oscillation. It is argued that phase detection and frequency modulation give the best results, with a higher signal-to-noise ratio (Porthun et al., 1998; Hartmann, 1999). However, these detection modes can require the addition of an electronics module to the microscope. In our MFM measurements we have used amplitude detection, which measures changes in the cantilever's amplitude of oscillation relative to the piezo drive. The signal depends on the force derivative in the following manner (Porthun et al., 1998):

$$f = f_0\sqrt{1 - \frac{\partial F / \partial z}{c}} \qquad (6)$$

with f_0 the free resonance frequency of the cantilever in the case of no tip sample interaction. In the amplitude detection, the cantilever is oscillated at a fixed frequency $f_{ext} > f_0$, where in the case of $\partial F / \partial z = 0$ the oscillation amplitude is already slightly below the maximum amplitude at f_0. When the resonance frequency changes this will result in a change in cantilever oscillation amplitude which can easily be detected. The disadvantage of this technique is that it is very slow for cantilevers with low damping and that a change in cantilever damping will be misinterpreted as change in resonance frequency.

It should be noted that an attractive interaction ($\partial F / \partial z > 0$) leads to a negative amplitude change (dark contrast in the image), while a repulsive interaction ($\partial F / \partial z < 0$) gives a positive amplitude variation (bright contrast).

Finally, Fig. 7(b) shows a typical MFM image. In this case, the sample was a piece of metal evaporated tape: a standard sample that is used to check whether the microscope is correctly tuned to image magnetic materials (Koch, 2005). It is clear that no correlation exists between the topography data shown on the left, and the magnetic data on the right. Consequently, the separation of both contributions is successful.

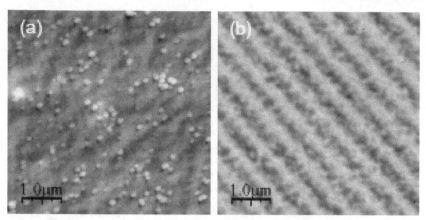

Fig. 7. Topographic image (a) and magnetic force gradient image (b) of a metal evaporated tape (Koch, 2005)

6. Applications of MFM in the study of Si and Ge-based magnetic semiconductors

6.1 Motivation

Driven by the promise of controlling charge and spin degrees of freedom, and its consequent technological impact through the realization of spintronic devices, many different ferromagnetic (FM) semiconductors have been investigated over the last few years. The potential advantages of this class of devices (in the form of ultra-dense non-volatile semiconductor memories, spin transistors and light emitting devices with polarized output, etc.) are expected to be, in addition to the low energy required to flip a spin: higher speed, greater efficiency and better stability (Zutic et al., 2004). Thus far, most of the work on FM semiconductors has been focused on Mn-containing II–VI or III–V compounds in which manganese replaces a fraction of group II or III sub-lattices (Dietl & Ohno, 2006). For practical reasons, however, the interest in a specific FM semiconductor depends on the existence of magnetic activity near or above room temperature as well as its compatibility with the current micro-electronics industry. Mn-containing Si- or Ge-based compounds partially fit these requirements since they possess a mature processing technology and because of some recent experimental work reporting Curie temperatures well above 300 K (Zhang et al., 2004; Kim et al., 2007). Furthermore, the low solubility of Mn in crystalline (c-)Si or c-Ge can be partially circumvented by using their amorphous counterparts, which also provide a more homogenous Mn distribution. Indeed, this is a particularly interesting feature since charge and spin states are sensitive mostly to the local environment so the magnetic activity existing in c-Si or c-Ge should also be observable in amorphous Si or Ge.

Based on these facts, this section reports on the MFM characterization of amorphous Si and Ge thin films containing different amounts of Mn and Co. Even though the amorphous character of the as-deposited films, thermal annealing at increasing temperatures induces their crystallization. Following this procedure, their magnetic properties have been systematically investigated as a function of the impurity concentration and atomic structure.

6.2 Experimental considerations

Thin films of amorphous SiMn and GeMn were prepared by conventional radio frequency sputtering. The Mn concentration ([Mn]) in the samples was in the ~ 0.1–24 at.% concentration range. Additionally, thin films of amorphous SiCo and GeCo were also deposited by sputtering. The Co concentration ([Co]) in the samples stayed in the ~ 1.7–10.3 at.% range. Pure samples were also prepared following identical conditions. The films, typically 1700 nm thick, were deposited principally on c-quartz and c-Si substrates. After deposition the films were submitted to thermal annealing treatments in the range of 200–900 ⁰C. The samples were characterized by a great variety of experimental techniques: (1) the composition of the films was determined mainly by energy dispersive x-ray spectrometry (EDS), (2) the atomic structure of the films was investigated by Raman scattering spectroscopy and x-ray diffraction (XRD) experiments, (3) the surface of the films was investigated by scanning electron microscopy (SEM) and AFM, (4) their optical properties were examined by means of transmission measurements, (5) the electrical resistivity of the films was measured using the standard van der Pauw technique, and (6) their magnetic properties were investigated by superconducting quantum interference device (SQUID) magnetometry and MFM. Except the SQUID measurements, all experimental characterizations were always carried out at room temperature. For further details, see Ferri et al., 2009a, 2009b, 2010a, 2010b, 2011.

6.3 Results and discussion

As confirmed by the Raman measurements, as the thermal annealing advances, the SiMn samples show crystallization signals that are accompanied by the growth of randomly dispersed sub-micrometre structures on the surface of the films. These structures are Mn-containing Si crystallites, surrounded by Si crystallites, amorphous Si and the $MnSi_{1.7}$ silicide phase (Ferri et al., 2009a). It is worth mentioning that the $MnSi_{1.7}$ is representative of a group of several Mn-silicides of the Mn_xSi_y form, with y/x approximately equal to 1.7: Mn_4Si_7, $Mn_{15}Si_{26}$, $Mn_{27}Si_{47}$, etc. Therefore, in this work, the Mn-silicides are simply identified by $MnSi_{1.7}$.

The morphology and magnetic characteristics of the $SiMn_{20\%}$ sample were investigated by means of AFM and MFM measurements (Fig. 8). Based on the AFM results the observed structures are typically ~ 750–1200 nm large and 300–400 nm high. Also, the image contrast present in Fig. 8(b) is a clear indication of the magnetic activity present in sample $SiMn_{20\%}$. At these dimensions, the contrast shown by the MFM images occurs because of force gradients between the FM tip and the magnetic activity present on the sample's surface. In this study, the MFM images were achieved after topography measurements (tapping mode) followed by sample surface scanning at a constant 200 nm height (lift mode). According to this procedure, no van der Waals forces are expected to be detected, and any change in the vibration amplitude of the cantilever is proportional to the gradient of magnetic fields perpendicular to the sample surface (Hartmann, 1999). It is worth noting that no MFM contrast was observed in the Mn-free film and $SiMn_{20\%}$ sample as-deposited nor after scanning the samples under the tapping mode.

In addition to the presence of magnetic activity in the sample under study, it also produces a remarkable contrast in the MFM image of Fig. 8(b). The FM materials are known to form

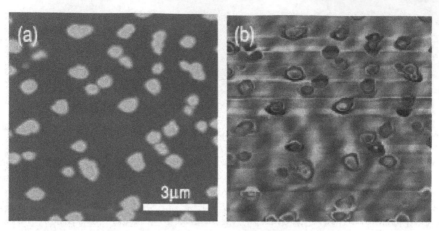

Fig. 8. (a) AFM and (b) MFM images of the sputter-deposited SiMn$_{20\%}$ film after thermal annealing at 600 °C. The AFM scanning was performed in the tapping mode, whereas MFM in the lift mode by means of a Co/Cr coated tip magnetized just before scanning. The measurements were carried out under room conditions (temperature and atmosphere) from a 1.7 μm thick film deposited on crystalline silicon (Ferri et al., 2009a)

domain structures to reduce their magnetostatic energy that, at very small dimensions such those experienced by a (sub-)micrometre dot, for example, adopts the configuration of a curling spin or magnetization vortex (Shinjo et al., 2000). When the dot thickness becomes much smaller than the dot diameter, all spins tend to align in-plane. In the curling configuration, the spin directions change gradually in-plane in order to maintain the exchange energy and to cancel the total dipole energy (Fig. 9). The development of these magnetic vortices is well documented in the literature and its comprehensive description can be found in many works (Zhu et al., 2002; Soares et al., 2008).

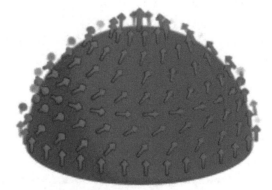

Fig. 9. Drawing of the magnetic moment configuration for ferromagnetic tri-dimensional sub-micrometre structures (Soares et al., 2008). At these very small dimensions, the magnetization adopts the pattern of a curling spin or magnetization vortex. In this curling arrangement, the spin directions change gradually in-plane in order to maintain the exchange energy and to cancel the total dipole energy

In this case, basically, the observed magnetic contrast occurs because of variations in the magnetization orientation along the sub-micrometre structures [Fig. 8(b)]. In other words, the presence of these Mn-based structures (probably Mn dimmers, in combination with the $MnSi_{1.7}$ phase) can lead to the appearance of magnetic activity (Bernardini et al., 2004; Affouda et al., 2006) whose main characteristics are highly influenced by the size and shape of the structures. Fig. 10 shows the surface topography in connection with the measured magnetic contrast of a single sub-micrometre structure. The figure also displays the height profile and MFM voltage achieved under horizontal [Fig. 10(b)], vertical [Fig. 10(c)] and diagonal [Fig. 10(d)] scans along the structure.

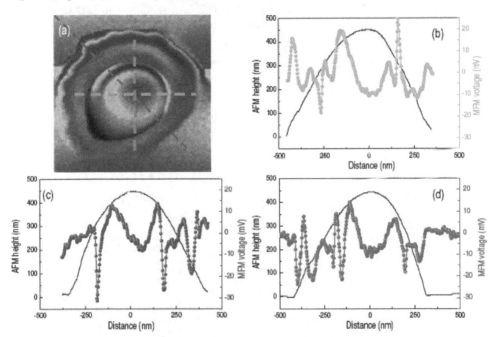

Fig. 10. (a) Magnetic force microscopy image of an isolated sub-micrometre structure present in the $SiMn_{20\%}$ film after thermal annealing at 600 °C. Its height profile (as obtained by AFM) and corresponding MFM voltage along the horizontal, vertical and diagonal dashed lines drawn in (a) are represented, respectively, in (b), (c) and (d). Note the MFM voltage pattern due to the presence of magnetic vortices in the structure (Ferri et al., 2009a)

It is interesting to observe the quite different topographic (AFM profile) and magnetic (MFM voltage) patterns achieved from the very same structure exclusively due to the presence of magnetic activity. The effect of manganese on the formation of these magnetic vortices is also remarkable suggesting that, once the structure is formed, the Mn distribution is non-uniform (and/or highly influenced by the presence of $MnSi_{1.7}$) around it.

The Mn-free, $GeMn_{3.7\%}$ and $GeMn_{24\%}$ films deposited under crystalline quartz substrates were also investigated through similar MFM measurements (Ferri et al., 2010a). Since these samples showed a flat surface, the magnetic activity of these three films was evaluated by scanning the MFM tip along a ~ 20 μm line across the crystalline quartz substrate partially covered by the desired Ge film (see sketch in Fig. 11). By adopting this procedure, at the bare substrate-film edge, the MFM tip will experience a signal difference which is proportional to the magnetic response of the probed region. Considering that crystalline quartz gives no magnetic contrast in the MFM measurements, the observed MFM signal is exclusively due to the GeMn films. In fact, and in accord with the literature (Cho et al, 2002) and our SQUID results, no MFM signal has been observed from both the amorphous and crystallized Mn-free Ge films. Also, and in order to confirm that the MFM signal is mainly of magnetic nature (Porthun et al., 1998), the measurements were carried out at a fixed tip-to-sample (substrate + film) distance d in the 100–2500 nm range. The main results of these MFM measurements, in conjunction with the SQUID data, are shown in Fig. 11. Here it is important to point out that similar results were obtained for the SiMn samples according this procedure (not shown).

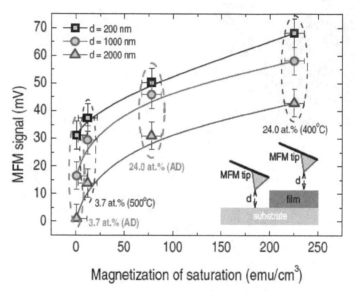

Fig. 11. MFM signal (as obtained from the voltage difference at the bare substrate-film edge region–see sketch) as a function of the magnetization of saturation (as obtained from the SQUID measurements at $T \geq 300$ K). The MFM data correspond to three different MFM tip-to-sample distances (d = 200, 1000, and 2000 nm). The measurements were carried out on the GeMn$_{3.7\%}$ and GeMn$_{24\%}$ films, deposited on crystalline quartz: both amorphous (AD–as-deposited) and after crystallization at the temperatures indicated in the figure. The lines joining the experimental data points are just guides to the eye (Ferri et al., 2010a)

The experimental data of Fig. 11 indicates that the MFM signal decreases with the distance d : demonstrating the magnetic character behind the interaction between the MFM tip and the sample. Except for minor deviations in the MFM signals obtained with the lowest d values, which were clearly affected by the experimental conditions (temperature, film thickness, and instrumental resolution, for example), the MFM signal scales with the magnetization of saturation, as obtained from the SQUID measurements. Indeed, the MFM signal increases with [Mn] and after the crystallization of the GeMn films. Therefore, as far as absolute magnetic data are available (such as those given by SQUID magnetometry, for example) the adopted experimental procedure can provide a convenient method to analyze the magnetic properties of microsized (or sub-microsized) isolated systems. As a final point, it is important to mention that the room temperature magnetic activity observed in the present GeMn samples (Fig. 11), occurs, basically, because of the presence of the Mn_5Ge_3 ferromagnetic germanide phase (Ferri et al., 2009b, 2010a).

For the magnetic characterization of the SiCo and GeCo films (deposited on crystalline quartz) the MFM technique was used similarly to the GeMn samples, since these samples also showed a flat surface (Ferri et al., 2010b). The main results of these MFM measurements are shown in Fig. 12, which illustrates results obtained in some SiCo and GeCo samples without annealing and after thermal treatment up to the crystallization temperature. In these samples, after crystallization, the non-magnetic $CoSi_2$ silicide and $CoGe_2$ germanide phases were found, as confirmed by XRD measurements (not shown). Therefore, we must to keep in mind that the only phase that can cause ferromagnetism at room temperature (or higher) for the samples in question, is the metallic Co, which has a Curie temperature of ~ 1382 K (Ko et al., 2006).

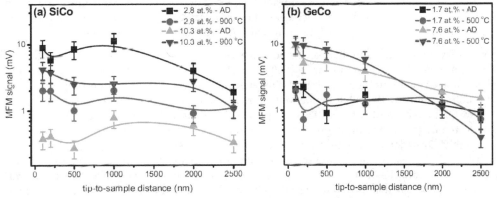

Fig. 12. MFM signal (as obtained from the voltage difference at the bare substrate-film edge region—see sketch of Fig. 11) as a function of the tip-to-sample distance, for as-deposited (AD) and thermally annealed (a) SiCo and (b) GeCo films (pure and containing different amounts of Co) deposited on c-quartz. The Co contents and the annealing temperatures are indicated in the figure. The lines joining the experimental data points are just guides to the eye

The MFM measurements for the Co-free Si and Ge films (both amorphous and annealed up to the crystallization temperature) suggest the absence of magnetic activity. This experimental result is expected [since it was also observed in the set of Mn-free Si and Ge samples (Ferri et al., 2009a, 2010)], and is in accord with the literature (Bolduc et al., 2005; Cho et al., 2002). When annealed at high temperatures, the XRD results indicate the presence of non-magnetic phases in the films containing Co. In addition, it is known that Co is less efficient than Mn in promoting ferromagnetic alignment, and a high magnetic moment, for the case of Ge (Continenza et al., 2006). Therefore, we expect a similar magnetic behaviour from the Co for the Si matrix. Taking these considerations into account, and remembering the fact that the MFM experiments were performed at room temperature, it is expected for the present samples a very weak or at least less intense magnetic signal than in the case of the Mn-containing films. Therefore, the results of Fig. 12 are in agreement with the initial expectations. Unlike observed in samples with Mn it is possible to identify only a slight decrease in the MFM signal with the tip-sample separation, due to the comparatively lower signal intensity. Here it is important to notice that the present procedure adopted in the MFM measurements is unique in the literature. Consequently, similar results obtained from others, for quantitative comparison purposes, are non-existent.

For the GeCo samples, we observed that the MFM signal intensity increased with increasing Co concentration [see Fig 12(b)]. The thermal treatment for samples with the same [Co], in principle, didn't intensify the magnetic signal. As an example of increasing MFM signal intensity with [Co], at a tip-sample separation of 500 nm, we observed that the Ge film with [Co] ~ 1.7 at.% showed a MFM signal of ~ 2 mV , and the Ge film with [Co] ~ 7.6 at.% exhibited a MFM signal of ~ 8 mV, both annealed at 500 °C. Still, as can be seen in Fig. 12(b), even the as-deposited Ge samples show magnetic signal, probably due to the existence of magnetically active Co atoms randomly distributed in the amorphous network. For the annealed samples, due to the diffusion of Co and the structural rearrangement of the network, it is expected that the number of magnetically active Co atoms increase (Ko et al., 2006). However, its magnetic activity does not exceed that of the amorphous films due to the formation of $CoGe_2$. Finally, the increasing in the magnetic signal with increasing [Co] is expected since the number of magnetically active Co atoms probably also increases.

For the SiCo films, the situation seems somewhat different, and not systematic. At first, as shown in Fig. 12(a), the sample with [Co] ~ 2.8 at.% without annealing shows a relatively high value of magnetic signal due to the magnetically active Co. After annealing at 900 °C, its value is diminished, probably due to the formation of $CoSi_2$. In contrast, the as-deposited Si film with the highest [Co] (~ 10.3 at.%) presents an extremely low MFM signal, probably due to the large number of magnetically inactive Co atoms, which may be associated with its highly disordered structure. After annealing at 900 °C, its magnetic activity is significantly increased due to the diffusion and consequent magnetic activation of the Co atoms. However, the magnetic activity is now limited by the existence of $CoSi_2$, and, therefore, its magnetism is less intense than the as-deposited film with [Co] ~ 2.8 at.%, that, in principle, doesn't have the silicide phase.

7. Conclusion

In summary, MFM is a relatively new technique for imaging magnetization patterns with high resolution and minimal sample preparation. The technique is an offspring of AFM and employs a sharp magnetic tip attached to a flexible cantilever. The tip is placed close to the sample surface (from some nanometres to a few micrometres) and interacts with the stray field emanating from the sample. The image is formed by scanning the tip laterally with respect to the sample and measuring the force (or force gradient) as a function of position. The interaction strength is then determined by monitoring the motion of the cantilever using a sensor. Although a lot of effort has been done in order to get quantitative information, MFM is still predominantly a qualitative characterization technique. In the present work, MFM proved to be particularly suitable to study the magnetic properties of Si and Ge-based magnetic semiconductors. In this context, the technique is very efficient to detect magnetic activity in the form of vortices in sub-micrometre structures. As well, a combination of the

MFM and SQUID techniques can be very convenient to probe the magnetic properties of microsized (or sub-microsized) isolated structures.

8. Acknowledgments

The authors are indebted to Professor Antonio Ricardo Zanatta (Instituto de Física de São Carlos, Universidade de São Paulo, Brazil) for the support with the deposition and characterization of the Si and Ge samples. This work was financially supported by the Brazilian agencies FAPESP and CNPq under CEPOF/INOF and INEO.

9. References

Affouda C. A., Bolduc M., Huang M. B., Ramos F., Dunn K. A., Thiel B., Agnello G., & LaBella V. P. (2006). Observation of crystallite formation in ferromagnetic Mn-implanted Si, *The Journal of Vacuum Science and Technology A*, Vol. 24, No. 4, pp. 1644-1647.

Bernardini F., Picozzi S., & Continenza A. (2004). Energetic stability and magnetic properties of Mn dimers in silicon, *Applied Physics Letters*, Vol. 84, No. 13, pp. 2289-2291.

Bolduc M., Affouda C. A., Stollenwerk A., Huang M. B., Ramos F. G., Agnello G., & Labella V. P. (2005). Above room temperature ferromagnetism in Mn-ion implanted Si, *Physical Review B*, Vol. 71, No. 3, pp. 033302-1–033302-4.

Bruker Corporation. (2011). MESP tips, In: *Bruker AFM Probes*, 23.09.2011, Available from: http://www.brukerafmprobes.com/Product.aspx?ProductID=3309.

Cho S., Choi S., Hong S. C., Kim Y., Ketterson J., Kim B. J., Kim Y. C., & Hung J. H. (2002). Ferromagnetism in Mn-doped Ge, *Physical Review B*, Vol. 66, No. 3, pp. 033303-1–033303-3.

Continenza A., Profeta G., & Picozzi S. (2006). Transition metal impurities in Ge: chemical trends and codoping studied by electronic structure calculations, *Physical Review B*, Vol. 73, No. 3, pp. 035212-1–035212-10.

Dietl T. & Ohno H. (2006). Engineering magnetism in semiconductors, *Materials Today*, Vol. 9, No. 11, pp. 18-26.

Ferri F. A. & Zanatta A. R. (2009). Structural, optical and morphological characterization of amorphous $Ge_{100-x}Mn_x$ films deposited by sputtering, *Journal of Physics D: Applied Physics*, Vol. 42, No. 3, pp. 035005-1–035005-6.

Ferri F. A. (2010). *Synthesis and characterization of Si and Ge based films doped with magnetic species*, PhD Thesis, Instituto de Física de São Carlos, Universidade de São Paulo, São Carlos, Brazil.

Ferri F. A., Pereira-da-Silva M. A., & Zanatta A. R. (2009). Evidence of magnetic vortices formation in Mn-based sub-micrometre structures embedded in Si–Mn films, *Journal of Physics D: Applied Physics*, Vol. 42, No. 13, pp. 132002-1–132002-5.

Ferri F. A., Pereira-da-Silva M. A., & Zanatta A. R. (2011). Development of the $MnSi_{1.7}$ phase in Mn-containing Si films, *Materials Chemistry and Physics*, Vol. 129, No. 1-2, pp. 148-153.

Ferri F. A., Pereira-da-Silva M. A., Zanatta A. R., Varella A. L. S., & de Oliveira A. J. A. (2010). Effect of Mn concentration and atomic structure on the magnetic properties of Ge thin films, *Journal of Applied Physics*, Vol. 108, No. 11, pp. 113922-1–113922-5.

Hartmann U. (1999). Magnetic force microscopy, *Annual Review of Materials Research*, Vol. 29, pp. 53-87.

Hendrych A., Kubínek R., & Zhukov A. V. (2007). The magnetic force microscopy and its capability for nanomagnetic studies - The short compendium, In: *Modern Research and Educational Topics in Microscopy*, Méndez-Vilas A. & Díaz J., (Eds.), Vol. 2, pp. 805-811, Formatex, ISBN 13: 978-84-611-9420-9, Badajoz, Spain.

Kim S. K., Cho Y. C., Jeong S. Y., Cho C. R., Park S. E., Lee J. H., Kim J. P., Kim Y. C., & Choi H. W. (2007). High-temperature ferromagnetism in amorphous semiconductor Ge_3Mn thin films, *Applied Physics Letters*, Vol. 90, No. 19, pp. 192505-1–192505-3.

Ko V., Teo K. L., Liew T., & Chong T. C. (2006). Ferromagnetism and anomalous Hall effect in Co_xGe_{1-x}, *Applied Physics Letters*, Vol. 89, No. 4, pp. 042504-1–042504-3.

Koch S. A. (2005). *Functionality and dynamics of deposited metal nanoclusters*, PhD Thesis, Groningen University Press, ISBN 90-367-2289-6, Groningen, The Netherlands.

Martin Y. & Wickramasinghe H. K. (1987). Magnetic imaging by "force microscopy" with 1000 Å resolution, *Applied Physics Letters*, Vol. 50, No. 20, pp. 1455-1457.

Porthun S., Abelmann L., & Lodder C. (1998). Magnetic force microscopy of thin film media for high density magnetic recording, *Journal of Magnetism and Magnetic Materials*, Vol. 182, No. 1-2, pp. 238-273.

Rasa M., Kuipers B. W. M., & Philipse A. P. (2002). Atomic force microscopy and magnetic force microscopy study of model colloids, *Journal of Colloid and Interface Science*, Vol. 250, No. 2, pp. 303-315.

Rugar D., Mamin H. J., Guethner P., Lambert S. E., Stern J. E., McFadyen I., & Yogi T. (1990). Magnetic force microscopy: General principles and application to longitudinal recording media, *Journal of Applied Physics*, Vol. 68, No. 3, pp. 11694-1184.

Schwarz A. & Wiesendanger R. (2008). Magnetic sensitive force microscopy, *Nanotoday*, Vol. 3, No. 1-2, pp. 28-39.

Shinjo T., Okuno T., Hassdorf R., Shigeto K., & Ono T. (2000). Magnetic vortex core observation in circular dots of permalloy, *Science*, Vol. 289, No. 5481, pp. 930-932.

Soares M. M., Biasi E., Coelho L. N., Santos M. C., Menezes F. S., Knobel M., Sampaio L. C., & Garcia F. (2008). Magnetic vortices in tridimensional nanomagnetic caps observed using transmission electron microscopy and magnetic force microscopy, *Physical Review B*, Vol. 77, No. 22, pp. 224405-1–224405-7.

Zhang F. M., Liu X. C., Gao J., Wu X. S., Du Y. W., Zhu H., Xiao J. Q., & Chen P. (2004). Investigation on the magnetic and electrical properties of crystalline $Mn_{0.05}Si_{0.95}$ films, *Applied Physics Letters*, Vol. 85, No. 5, pp. 786-788.

Zhu X., Grütter P., Metlushko V., & Ilic B. (2002). Magnetization reversal and configurational anisotropy of dense permalloy dot arrays, *Applied Physics Letters*, Vol. 80, No. 25, pp. 4789-4791.

Zutic I., Fabian J., & DasSarma S. (2004). Spintronics: Fundamentals and applications, *Reviews of Modern Physics*, Vol. 76, No. 2, pp. 323-410.

3

Crystal Lattice Imaging Using Atomic Force Microscopy

Vishal Gupta

FLSmidth Salt Lake City Inc.

USA

1. Introduction

Atomic force microscopy (AFM) has been a very useful tool in interrogating the micron-to-nano sized structures at both atomic and subnanometer resolution. AFM allows both imaging of surfaces and interactions with surfaces of interest to help researchers explain the crystal lattice structure, and surface chemical and mechanical properties at nano scale. Since the invention of AFM, one has been frequently attracted by AFM images when browsing through many scientific publications in physics, chemistry, materials, geology, and biology (Gan, 2009; Sokolov et al., 1999; Wicks et al., 1994). AFM has been successfully used for imaging solid surfaces with subnanometer resolution for natural materials such as minerals, synthetic materials such as polymers and ceramics, and biological materials such as live organisms. There are also numerous reports of molecular and subnanometer resolution on biological and polymer samples.

Atomic force microscopy (AFM) has been quite successfully used by scientists and researchers in obtaining the atomic resolution images of mineral surfaces. It is quite amazing to see the individual atoms, and their arrangements, that make up the surfaces. In some cases, atoms from the mineral surfaces can be deliberately removed with the AFM so that the internal structure of the surface can be studied.

The key to obtaining atomic-scale imaging is precisely control the interactions between the atoms of the scanning tip and the atoms of the surface being studied. Ideally a single atom of the tip is attracted or repelled by successive atoms of the surface being studied. However, this is a dynamic environment and there can be accidental or deliberate wear of the tip and the surface, so the situation is far from ideal. A number of theoretical and practical studies have added some understanding of this interaction but our understanding is still incomplete (Nagy, 1994). Despite the imperfect knowledge, application of the instrument to mineral studies demonstrates that the AFM works well, often at atomic scale resolution.

It is now well established with some success that AFM can also be used to investigate the crystal lattice structure of mineral surfaces. Atomic resolution has been successfully obtained on graphite (Albrecht & Quate, 1988; Sugawara et al., 1991), molybdenum sulfide (Albrecht & Quate, 1988), boron nitride (Albrecht & Quate, 1987), germanium (Gould et al., 1990), sapphire (Gan et al., 2007), albite (Drake & Hellmann, 1991), calcite (Ohnesorge & Binnig, 1993) and sodium chloride (Meyer & Amer, 1990). The AFM has also been used to investigate the crystal lattice structure of the tetrahedral layer of clay minerals in 2:1 layer structures, such as muscovite (Drake *et al.*, 1989), illite (Hartman *et al.*, 1990) and montmorillonite (Hartman *et al.*, 1990). Atomic-scale resolution has also been obtained for the basal oxygen atoms of a mixed-layered illite/smectite (Lindgreen *et al.*, 1992), zeolite clinoptilolite (Gould *et al.*, 1990) and hematite (Johnsson *et al.*, 1991).

Wicks *et al.* (Wicks *et al.*, 1992) were probably the first to simultaneously report the surface images of both the tetrahedral and octahedral sheets of lizardite (1:1 layer structure) using AFM, and they identified the surface hydroxyl groups and magnesium atoms in the octahedral sheet. In this way, they identified the two sides of the lizardite clay mineral. The surface images of chlorite (2:1:1-type structure) were also investigated by AFM, and both the tetrahedral sheet and the brucite-like interlayer sheet were observed (Vrdoljak *et al.*, 1994). Recently, Kumai *et al.* (Kumai *et al.*, 1995) examined the kaolinite surface using AFM. They used the "pressed" powder sample preparation technique, and obtained the surface images of both the silica tetrahedral surface and alumina octahedral surface of kaolinite particles.

Despite great success in obtaining atomic resolution, AFM images may be subject to various distortions, such as instrumental noise, drift of the piezo, calibration problem with the piezo, vibrations, thermal fluctuations, artifacts created by the AFM tip, contamination of the mineral or tip surface, and tip induced surface deformations. The initial AFM images are often noisy and suffer from instrumental effects such as image bow due to sample tilt. Some of these problems can be fixed with data processing software by applying appropriate flattening, filter out low frequency noise, and clarify the structural details in an image using two dimensional fast-Fourier transforms (2DFFT). Some of these fixtures will be discussed in the subsequent section of this chapter by following a case study on obtaining crystal lattice images of kaolinite. Similar experimental routine could be applied on obtaining atomic resolution images of any surface of interest.

This chapter summarizes the achievement of AFM to obtain atomic resolution images of mineral surfaces. In particular, a case study for obtaining crystal lattice images on kaolinite surface will be presented. The principles of AFM and its different modes of operation will be introduced. A brief introduction of image acquisition and filtering routines will be discussed followed by tip and surface interaction. This will be followed by different ways to acquire images with atomic resolution. The important issues of reproducibility and artifacts will be discussed. A critical review of literature will be supplemented in each section for obtaining atomic resolution images. Finally, the new challenges for AFM to obtain atomic resolution images on the complex surfaces will be discussed.

2. Basic principles and operation modes of AFM

2.1 Principles of AFM

An AFM consists of a probe, scanner, controller, and signal processing unit-computer. AFM works by rastering a sharp probe across the surface to obtain a three-dimensional surface topograph. As the probe rasters, it feels the highs and lows of surface topography through complex mechanisms of tip-surface interactions. These signals are sent back via a laser reflected back from the probe surface to a photo-detector. The photo-detector through a feedback control loop, keep the tip at constant height or constant force from the surface. The feedback signals are sent to a signal processing software, which generates a three-dimensional topograph of the surface.

2.2 Operation modes of AFM

The operating modes of AFM can be divided *into static* (*DC*) *mode* – the probe does not vibrate during imaging, and *dynamic* (*AC*) *mode* – the cantilever is excited to vibrate at or off its resonant frequency. The dynamic mode AFM can be either an amplitude-modulated AFM (AM-AFM) or a frequency-modulated AFM (FM-AFM). Usually, AM-AFM is referred to as intermittent contact mode or tapping mode. The imaging could be conducted by manipulating the repulsive interaction between a probe and the surface, which is referred to as contact mode imaging. When the probe images the surface with an attractive interaction, is usually referred to as non-contact mode. Note that both the DC and AC modes may be operated in contact mode; in most cases, however, DC mode is referred to as contact mode. FM-AFM is usually referred as non-contact mode.

2.2.1 Contact mode

The contact mode can be operated in constant force mode or constant height mode, depending on whether the feedback loop is turned on. *Constant force* requires a setpoint that needs to be manually adjusted to compensate for the drift during imaging or to control the tip-surface force. This is done on non-atomically smooth surface. The piezo-drive signal is used for generating the height signal on a topograph. *Constant height* mode is most suitable for scanning atomically smooth surfaces at a fixed setpoint (tip-surface force). The deflection of the cantilever is used for generating the height signal on a topograph.

2.2.2 Intermittent or tapping mode

The intermittent or tapping mode (or AM-AFM) is usually conducted on soft samples, such as loosely attached structure on the surface or even more delicate biological samples such as DNA, cells and micro-organisms. The probe is excited at a setpoint amplitude of cantilever oscillation. The amplitude of the cantilever dampens from full oscillation (non contact) to smaller oscillations when it encounters a structure on the surface (intermittent contact). The change in the amplitude of the probe stores the structural information of the surface, which generates a three-dimensional topograph. A large setpoint amplitude is required in the noncontact region, and a small setpoint amplitude is required in the intermittent contact regime. For example, Gan (Gan, 2009) pointed out that the Magonov group achieved molecular resolution with AM-AFM (Belikov & Magonov, 2006; Klinov & Magonov, 2004) and the Engel group achieved subnanometer resolution on protein samples (Moller *et al.*, 1999).

2.2.3 Non-contact mode

Martin *et al.* (Martin *et al.*, 1987) introduced the concept of non-contact mode (FM-AFM) in 1987 to precisely measure the interaction force between a probe and the surface. During non-contact mode, the probe is excited to oscillate at its resonant frequency. The frequency shift of a probe is monitored, as it encounters a surface structure, which generates surface topograph. Giessibl (Giessibl, 2000) was able to use an AFM in non-contact mode to obtain atomic resolution images of reactive surfaces such as Si.

3. Image acquisition and filtering

The quality of the raw data is of primary importance in obtaining high resolution AFM images. A good quality raw image must be obtained without the use of online filters. Special attention should be taken in order to determine if an image is of good quality for a particular sample and whether the image is real or not. This is done by varying scan direction and speed, varying instrumental gains and contact force, changing sample locations, retracting and extending the tip, and collecting multiple set of data from different sample and using different tips. Other factors such as varying color contrast/offset, z-height range, and checking for periodicity, by looking at the screen close up and from a distance, are important.

Once a good raw image has been obtained, some filtering can be applied to enhance the features seen in the image, and to distinguish between instrumental artifacts and real features. A filtered image should always be compared to the unfiltered image as a cross check to ensure that artifacts have not been introduced as a result of filtering.

A variety of data processing programs are available to filter images. These will vary from instrument to instrument and are explicitly described in the user's manual. Some of the commonly followed filter routines are described below.

3.1 Flattening

Flattening subtracts the average value of the height of each scan line from each point in the scan line and reduces the effect of image bow and vibration in the Y direction. This could be applied automatically during real time imaging or manually after the image is captured. At times, a plane is fitted to the captured image. Plane fit calculates a best fit second order polynomial plane, and subtracts it from the image. Usually this is applied once in X direction and once in Y direction

3.2 Low pass / high pass filters

Lowpass filtering replaces each data point in the image with a weighted average of the 3 x 3 cell of points surrounding and including the point. It may be applied a number of times. This removes the high frequency noise, but it also reduces image resolution by "defocusing" the periodic features observed in the raw data. Highpass filtering on the other hand replaces each data point with a weighted difference between the data point and each of its eight neighbors. This routine is particularly good for enhancing height differences within an image.

3.3 2-Dimensional Fast Fourier Transforms (2DFFT)

This is the most useful filtering routine which can greatly improve images. This technique converts the image to the frequency domain by calculating the 2-dimensional power spectrum or 2DFFT. The 2DFFT of the image may then be filtered and an inverse transform performed on the filtered data to produce a new image. This routine should be practiced with care by resizing the image to the maximum pixel dimensions, prior to the application of the 2DFFT, and then varying color contrast/offset of the power spectrum image. Some criticism of this technique by AFM users were reported as (1) 2DFFT may introduce the features which are not present in the initial image, and (2) use of a 2DFFT smears the atomic positions so that the resolution of individual atom is not obtained. In first case, it's possible to introduce the artifacts after 2DFFT processing, and it's a matter of experience and competence in selecting or rejecting the right periodicities to obtain an image. In contradiction to the second criticism, Wicks *et al.* (Wicks *et al.*, 1994) successfully reported two different atomic–repeat units of lizardite in a single image during a high tracking force experiments. They demonstrated that this criticism of 2DFFT is not valid.

4. Resolution

AFM is a computer-controlled local probe technique which makes it difficult to give a straightforward definition of resolution. The AFM vertical resolution is mainly limited by thermal noise of the deflection detection system. Most commercial AFM instruments can reach a vertical resolution as low as 0.01 nm for more rigid cantilevers. The lateral resolution of AFM is defined as the minimum detectable distance between two sharp spikes of different heights. A sharp tip is critical for achieving high resolution images. Readers may refer to Gan (Gan, 2009) for more discussion on probe sizes.

Despite great success by researchers in obtaining atomic resolution images, AFM is looked at with doubt as compared to scanning tunneling microscopy. These doubts about resolution have been dispersed. For example, Ohnesorge and Binnig (Ohnesorge & Binnig, 1993) obtained images of the oxygen atoms standing out from the cleavage plane of calcite surface in water. Similarly, Wicks *et al.* (Wicks *et al.*, 1993) used high tracking force to strip away the oxygen and silicon of the tetrahedral sheet to image the interior O, OH plane of lizardite at atomic resolution. Recently, Gupta *et al.* (Gupta *et al.*, 2010) showed high resolution images of silica tetrahedral layer and alumina octahedral layer of kaolinite surface.

5. Tip-surface interaction

Tip-surface forces are of paramount importance for achieving high resolution AFM images. They can be described based on (i) continuum mechanics, (ii) the long range van der Waals force, (iii) the capillary force, (iv) the short range forces, (v) the electrical double layer force in a liquid, and (vi) contamination effects.

A continuum model treats the materials of the tip and sample as continuum solids. Various continuum models such the Hertz model, the JKR model, the MD model, and the Schwarz model consider mechanical deformation or surface energy alone or both. At high applied force, the tip and the substrate may deform inelastically. One should thus be cautious in using continuum models to predict tip-surface interactions. The van der Waals (vdW) force between macroscopic objects is due to the dispersion interactions of a large number of atoms between two objects interacting across a medium. The strength of the vdW force is measured with the Hamaker constant. The macroscopic vdW force is determined by the properties of the materials and the medium, and the tip geometry. In most cases, vdW forces are attractive between tip and surface of interest. The capillary force arises when tip approaches the surface in air. The water molecules on the surface forms a bridge with the tip and an increased force must be applied in order to detach the tip from the surface. This increased force is called the capillary force, and depends on the surface properties, humidity, temperature and geometry of the tip. The capillary force is usually more long-ranged than the van der Wall force under moderate humidity conditions. Short-range forces become important when the tip-surface distance is less than 1 nm. Short-range force may originate from Born repulsion, chemical bonding, and electrostatic and vdW interactions between atoms. The electrical double layer force arises when two surfaces approach each other in solution. The surfaces develop charges either by protonation/de-protonation, adsorption, and specific chemical interaction, which attracts counterions and co-ions from solution. Lastly, contaminants, particularly organic materials adheres either to surface or tip, even in trace amounts, can significantly affect the tip-surface interaction. Therefore, a clean tip and surface are highly desirable prior to and throughout the experiments. This is a brief review of tip-surface interactions, and readers are advised to review classic textbooks (Butt *et al.*, 2003; Israelachvili, 1985; Masliyah & Bhattacharjee, 2006).

In order to achieve atomic resolution image, the external load on the tip must counteract the tip-surface interactions discussed above. The external load is a function of spring constant of the tip and its bending. It is highly desirable to keep the tip load as low as possible to produce high resolution image.

6. General tips to achieve atomic resolution

Atomic resolution images can be obtained by controlling tip-surface interactions as discussed above. In addition to tip-surface interactions, the following suggestions can be made to achieve atomic resolution:

- Use sharper tips. Weih *et al.* (Weihs *et al.*, 1991) showed by calculations that the lateral resolution increased by a factor of 4 by reducing the tip radius from 200 to 20 nm. The sharper tip also reduces the adhesion force between a tip and the surface, which also decrease the tip load.
- Use stiffer tips. The elastic modulus could be increased by using a stiffer tip to achieve a smaller contact area between a tip and the surface. The smaller contact area between a tip and the surface is desired so as to realize only few atoms in contact. Ideally, a single atom of the tip should interact with each surface atom to obtain atomic resolution.
- Reduce tip load. By applying the cantilever bending force, the contact area between a tip and the surface could be reduced by lowering the tip load.
- Reduce adhesion. The work of adhesion can be minimized by immersing the tip and sample in liquid.

7. Artifacts and reproducibility

The topographs obtained by AFM should be reproducible and represent the real surface structure of the sample. Artifacts at the atomic scale are topographic features by which uncertainties and errors enter the surface structure determination. There are numerous types of AFM artifacts, including missing atoms/molecules/vacancies, ghost atoms, and fuzzy steps etc. Most artifacts are caused by multiple-tip surface contacts and high tip loads. Ideally, a single atom tip interacts with the surface to obtain atomically resolved topographs. In reality, however, the structure, geometry, and surface chemistry of the AFM tips are usually poorly defined. During imaging, the AFM tip may get deformed and cause multiple point contacts. It is therefore highly desirable to monitor the structural and chemical modification of the tip before and after experiments. Equally, the low tip load is desirable for achieving high resolution atomic images. Ohnesorge and Binnig (Ohnesorge & Binnig, 1993) have demonstrated the dramatic change in topograph by carefully controlling the tip-surface interaction. Sokolov and Henderson (Sokolov & Henderson, 2000) also showed that an increased tip load destroys the atomically resolved images determined from the vertical force contrast and only improves lattice resolution images determined from the friction forces. Cleveland et al. (Cleveland et al., 1995) also showed through atomic imaging of calcite and mica surfaces in water, that surface atoms could only be unambiguously identified when the tip load was attractive. It is thus highly recommended that one be cautious in interpreting AFM images before systematic studies of the tip load effect are carried out.

The AFM images should show the real surface structure and be reproducible. The surface structure should remain unchanged with varying probes, scanning directions, different location on the same surface, different sample of same material, tip-surface forces, and even different instruments and techniques if possible.

Finally, more confidence in the recorded AFM topographs will be gained if the same surface can be analyzed with other techniques such as STM, high resolution transmission electron microscopy, x-ray crystallography etc. Electron microscopy requires complex surface preparation procedures, but they are free from artifacts introduced in AFM images. These alternative techniques may compliment AFM in obtaining and verifying the atomic images.

8. Case study: Crystal lattice imaging of silica face and alumina face of kaolinite

Kaolinite naturally exists as pseudo-hexagonal, platy-shaped, thin particles generally having a size of less than one micron extending down to 100 nm. The crystallographic structure of kaolinite suggests that there should be two types of surface faces defined by the 001 and the 001 basal planes. In this way, one face should be described by a silica tetrahedral surface and the other face should be described by an aluminum hydroxide (alumina) octahedral surface as shown in figure 1. The objective of this case study is to demonstrate the bi-layer structure of kaolinite – a silica tetrahedral layer and an alumina octahedral layer, through atomic resolution obtained using AFM.

8.1 Materials and methods

8.1.1 Sample preparation

A clean English kaolin (Imerys Inc., UK) was obtained from the St. Austell area in Cornwall, UK. The sample was cleaned with water and elutriation was used to achieve classification at a size of less than 2 μm. No other chemical treatment was done. Further details about the kaolinite extraction and preparation are given in the literature (Bidwell et al., 1970).

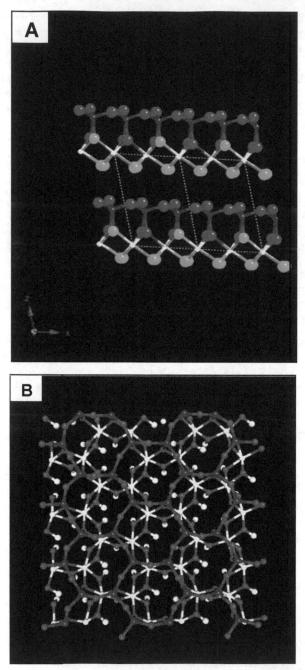

Fig. 1. Side view (A) and Top view (B) of kaolinite (001) surface structure. The silica tetrahedra (red: oxygen, blue: silicon) and alumina octahedra (yellow: aluminum, green: hydroxyl) bilayers thought to be bound together via hydrogen bonding are illustrated in (A).

The kaolinite suspension (1000 ppm) was prepared in high purity Milli-Q water (Millipore Inc.) with a resistivity of 18.2 MΩ-cm. The pH was adjusted to 5.5 using 0.1 M HCl or 0.1 M KOH solutions.

8.1.2 Substrate preparation

Two substrates – a mica disc (ProSciTech, Queensland, Australia) and a fused alumina substrate (Red Optronics, Mountain View, CA), were used to order the kaolinite particles (Gupta & Miller, 2010). The kaolinite particle suspension (1000 ppm) was sonicated for 2 minutes, and about 10 µl of the suspension was air-dried overnight on a freshly cleaved mica substrate under a petri-dish cover in a laminar-flow fume hood. In this way, the kaolinite particles attach to the mica substrate with the alumina face down exposing the silica face of kaolinite, as shown from previous surface force measurements (Gupta & Miller, 2010), i.e., the positively charged alumina face of kaolinite is attached to the negatively charged mica substrate.

The fused alumina substrate was cleaned using piranha solution (a mixture of sulfuric acid and hydrogen peroxide in a ratio of 3:1) at 120⁰C for 15 minutes, followed by rinsing with copious amounts of Milli-Q water, and finally blown dry with ultra high purity N_2 gas. A 10 µl kaolinite suspension was applied to the alumina substrate and dried in the same manner as the mica. It was found that the alumina face of kaolinite was exposed on the fused alumina substrate based on previous surface force measurements (Gupta & Miller, 2010), i.e., the negatively charged silica face of kaolinite is attached to the positively charged fused-alumina substrate.

The samples were prepared the night before AFM analysis and stored in a desiccator until their use. Just prior to the AFM experiments, the substrates were sonicated for a minute in Milli-Q water to remove loosely adhered kaolinite particles, washed with Milli-Q water, and gently blown with N_2 gas before AFM investigation. All substrates were attached to a standard sample puck using double-sided tape.

8.1.3 Atomic Force Microscopy

A Nanoscope AFM with Nanoscope IV controller (Veeco Instruments Inc., Santa Barbara, CA) was used with an E-type scanner. Triangular beam silicon nitride (Si_3N_4) cantilevers (Veeco Instruments Inc., Santa Barbara, CA), having pyramid-shaped tips with spring constants of about 0.58 N/m, were used. The cantilevers were cleaned using acetone, ethanol, water in that order, and gently dried with ultra high purity N_2 gas. The cantilevers were subsequently cleaned in a UV chamber for 30 minutes prior to use. The substrates were loaded on AFM equipped with a fluid cell. The contact mode imaging was done in Milli-Q water. The AFM instrument was kept in an acoustic and vibration isolation chamber. The imaging was commenced 30 minutes after sample loading to allow the thermal vibration of the cantilever to equilibrate in the fluid cell. First, an image of the particles was obtained at a scan rate of 1 Hz and scan area of 1 µm. Subsequently, the atomic resolution imaging was completed using the zoom-in and offset feature of the Nanoscope vs. 5.31R1 software (Veeco Instruments Inc., Santa Barbara, CA) to scan an area of 12 nm on the particle surface. The atomic imaging was obtained at a scan rate of 30 Hz at scan angle of 80⁰–90⁰ with very low integral and proportional gain (0.06). The online filters (low pass and high pass) were turned off during the online crystal lattice imaging.

During offline image processing, flattening and low pass filtering were applied to obtain clear images using Nanoscope vs. 5.31R1 software. The images were further Fourier-filtered (2D FFT) to obtain the crystal lattice images using SPIP software (Image Metrology A/S, Denmark).

8.2 Results and discussion

In order to obtain the crystal lattice imaging of the silica face and alumina face of kaolinite, the scanner was first calibrated using a mica substrate. Figure 8.2 presents the crystal lattice imaging of mica, which shows the height image, fast-Fourier transform (FFT) spectra and the FFT transformed height image. In order to make sure that the image is real, the imaging was acquired from other locations on the mica substrate and also with varying scan size and scan angle. The repeated pattern of dark and light spots was reproducible and the dark spots observed were scaled appropriately with the scan size and angle. The images showed some drift in both x and y direction during imaging. The dark spots in Figures 8.2C and 8.2D correspond to a hole surrounded by the hexagonal lattice of oxygen atoms. The light spots are attributed to the three-surface oxygen atoms forming a SiO_4 tetrahedron or pairs of SiO_4 tetrahedra forming a hexagonal ring-like network. Similar images were reported for the 1:1 type clay mineral, lizardite (Wicks *et al.*, 1992) and other 2:1 type clay minerals (Drake *et al.*, 1989; Hartman *et al.*, 1990; Sharp *et al.*, 1993) from AFM observations on a single crystal. The fast-Fourier transform showed the intensity peaks of oxygen atoms arranged in a hexagonal ring network (see Figure 8.2B). The crystal lattice spacing between neighboring oxygen atoms was calculated as 0.51 ± 0.08 nm, from the average of 10 neighboring atoms. This is in very good agreement with the literature value of 0.519 nm (Wicks *et al.*, 1993).

Figure 8.3 shows an image of a kaolinite particle on a mica substrate. The image shows the platy nature and the pseudo-hexagonal shape of the kaolinite particle. The scanning was sequentially zoomed on the particle. Figure 8.4 shows the crystal lattice imaging of the silica face of a kaolinite particle on the mica substrate. The flattening and low pass filtering was applied to the height image in an offline mode (see Figure 8.4B). The FFT spectra showed the similar intensity of peaks of oxygen atoms arranged in a hexagonal ring network as observed for the mica substrate. As expected, the silica face of kaolinite showed the similar hexagonal ring-like network of oxygen atoms as observed on the mica substrate (compare Figure 8.2D and Figure 8.4D). Note that the scan scale for the image of the silica face of kaolinite was twice that used for the mica substrate (12 nm vs. 6 nm), which shows the reproducibility of the crystal lattice images obtained on different substrates. The crystal lattice spacing between neighboring oxygen atoms was calculated as 0.50 ± 0.04 nm, from the average of 10 neighboring atoms. This lattice spacing is in good agreement with 0.53 nm as reported in the literature (Kumai *et al.*, 1995).

The crystal lattice imaging of the alumina face of kaolinite on a fused alumina substrate is shown in Figure 8.5. The FFT spectra shows the intensity peaks of the hydroxyl atoms forming a hexagonal ring network similar to that obtained on the silica face of kaolinite (see Figure 8.5C). Notice that the hexagonal ring of hydroxyls shows the inner hydroxyl in the center of the ring instead of a hole as observed for the silica face of kaolinite and mica

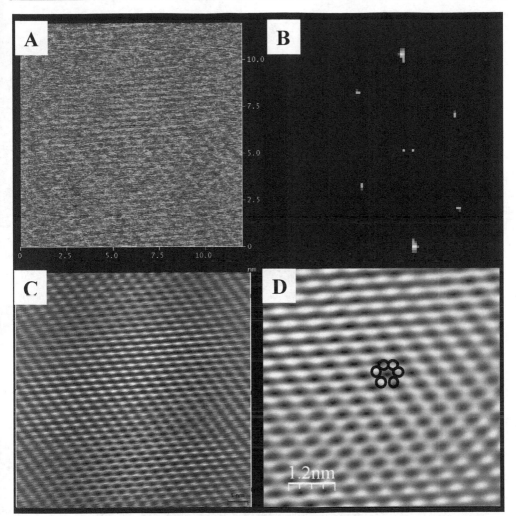

Fig. 2. Crystal lattice imaging of mica substrate showing (A) Flattened height image, (B) FFT spectra, (C) FFT transformed flattened height image, and (D) Zoomed-in image of (C) of scan area of 36 nm^2. The six light spots in (D) show the hexagonal ring of oxygen atoms around the dark spots representing a hole. Adapted from (Gupta *et al.*, 2010).

substrates (compare Figure 8.2D, Figure 8.4D, and Figure 8.5D). The image shown in Figure 8.5D is similar to the octahedral sheet of lizardite (Wicks *et al.*, 1992), the internal octahedral sheets of micas and chlorite (Wicks *et al.*, 1993), and the brucite-like layers of hydrotalcite (Cai *et al.*, 1994). The octahedral sheet of kaolinite consists of a plane of hydroxyls on the surface. The average hydroxyl-hydroxyl distance of the octahedral sheet is 0.36 ± 0.04 nm which is in reasonable agreement with the literature value of 0.29 nm (Wyckoff, 1968). For a kaolinite pellet, Kumai *et al.* (Kumai *et al.*, 1995) observed the distance between the hydroxyl atoms as 0.33 nm.

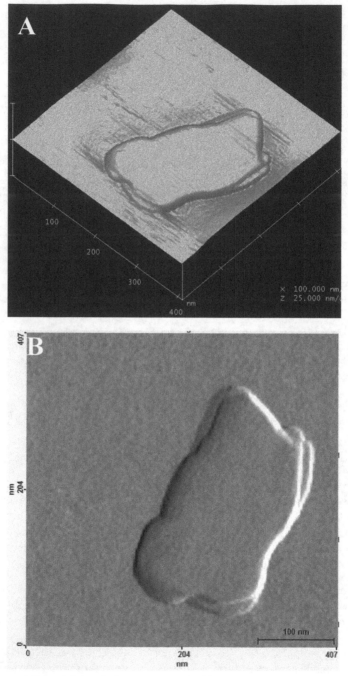

Fig. 3. (A) Topography, and (B) Deflection images of kaolinite particle on the mica substrate. Adapted from (Gupta *et al.*, 2010).

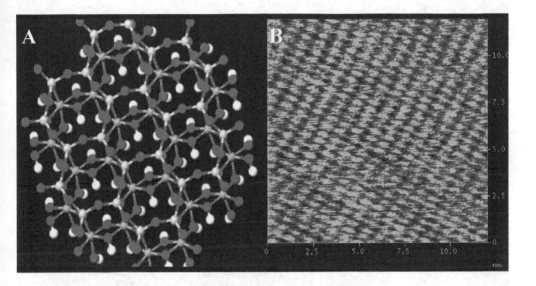

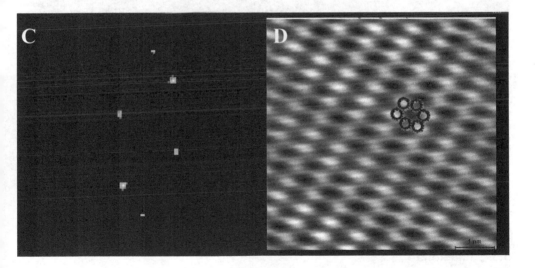

Fig. 4. Crystal lattice imaging of the silica face of kaolinite showing (A) Theoretical atomic lattice structure, (B) Flattened-low pass filtered height image, (C) FFT spectra, and (D) FFT transformed flattened-low pass filtered height image of scan size 36 nm². The six black circles in (D) show the hexagonal ring of oxygen atoms around the dark spots representing a hole. Adapted from (Gupta et al., 2010).

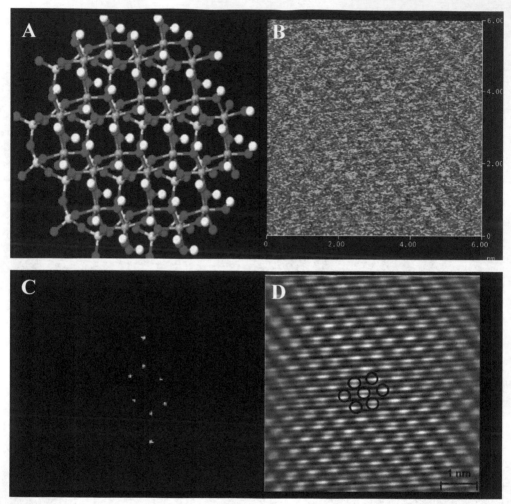

Fig. 5. Crystal lattice imaging of the alumina face of kaolinite showing (A) Theoretical atomic lattice structure, (B) Flattened-low pass filtered height image, (C) FFT spectra, and (D) FFT transformed flattened-low pass filtered height image of (B). The seven black circles in (D) show the hexagonal ring of hydroxyl atoms with a central inner hydroxyl atom. Adapted from (Gupta *et al.*, 2010).

9. Conclusions

For the last two decades, AFM has been established as an important tool for the study of surfaces. AFM produces information with minimal surface preparation that is not matched by other techniques. The quality of images has increased, as our understanding of the theory of the interaction of the tip and the sample. Atomic resolution images recorded on a variety of samples such as natural minerals, synthetic materials, zeolites, biological samples etc. have established the AFM as the microscope for the atomic scale.

Looking ahead, we must face several challenges to produce fast and reproducible atomic resolution images. One should be skeptical of high resolution topographs, and do diligent work in reporting data. The image acquisition procedures and filtering routines discussed in this chapter should be applied judiciously. One should be aware of artifacts introduced during real-time image acquisition or post processing should be dealt with cautiously, and must be reported. Probes play a key role in realizing high resolution topographs. The benefits of sharper tips are numerous, such as smaller contact area and reduced long range forces. Most conventional tips are made from silicon nitride and silicon. Polymers or diamond tips have also used in some applications (Beuret *et al.*, 2000). Recent developments in producing nano-tips through whiskers or carbon fiber may find potential application in AFM for high resolution images (Marcus *et al.*, 1989; Marcus *et al.*, 1990).

Recent developments on cantilever dynamic studies (Holscher *et al.*, 2006; Strus *et al.*, 2005) and new experimental techniques, such as Q-control (Ebeling *et al.*, 2006; Okajima *et al.*, 2003) and higher order vibration imaging (Martinez *et al.*, 2006) will very likely make AFM a powerful tool for high resolution characterization in the future. Despite recent developments in AFM instrumentation for precise control of tip movement, it is still highly desirable to confirm the reliability of AFM topographs with complimentary techniques such as transmission electron microscopy (Matsko, 2007). We can conclude that AFM is a powerful instrument, and could be used for studying a variety of surfaces.

10. References

Albrecht, T. R. and Quate, C. F. (1987). Atomic Resolution Imaging of a Nonconductor by Atomic Force Microscopy. *Journal of Applied Physics*, Vol. 62, No. 7, pp. (2599-2602), 0021-8979

Albrecht, T. R. and Quate, C. F. (1988). Atomic Resolution with the Atomic Force Microscope on Conductors and Nonconductors. *Journal of Vacuum Science & Technology , A*, Vol. 6, No. 2, pp. (271-274), 0734-2101

Belikov, S. and Magonov, S. (2006). True Molecular-Scale Imaging in Atomic Force Microscopy: Experiment and Modeling. *Japanese Journal of Applied Physics, Part 1 (Regular Papers, Short Notes & Review Papers)*, Vol. 45, No. 3B, pp. (2158-2165), 0021-4922

Beuret, C., Akiyama, T., Staufer, U., De, R. N. F., Niedermann, P. and Hanni, W. (2000). Conical Diamond Tips Realized by a Double-Molding Process for High-Resolution Profilometry and Atomic Force Microscopy Applications. *Applied Physics Letters*, Vol. 76, No. 12, pp. (1621-1623), 00036951

Bidwell, J. I., Jepson, W. B. and Toms, G. L. (1970). The Interaction of Kaolinite with Polyphosphate and Polyacrylate in Aqueous Solutions - Some Preliminary Results. *Clay Minerals*, Vol. 8, No. pp. (445–459)

Butt, H. J., Graf, K. and Kappl, M. (2003). *Physics and Chemistry of Interfaces* Wiley-VCH

Cai, H., Hillier, A. C., Franklin, K. R., Nunn, C. C. and Ward, M. D. (1994). Nanoscale Imaging of Molecular Adsorption. *Science (Washington, D. C.)*, Vol. 266, No. 5190, pp. (1551-1555)

Cleveland, J. P., Radmacher, M. and Hansma, P. K. (1995). Atomic-Scale Force Mapping with Atomic Force Microscope. *NATO ASI Ser., Ser. E*, Vol. 286, No. Copyright (C) 2010 American Chemical Society (ACS). pp. (543-549), 0168-132X

Drake, B. and Hellmann, R. (1991). Atomic Force Microscopy Imaging of the Albite (010) Surface. *American Mineralogist*, Vol. 76, No. 9-10, pp. (1773-1776)

Drake, B., Prater, C. B., Weisenhorn, A. L., Gould, S. A., Albrecht, T. R., Quate, C. F., Cannell, D. S., Hansma, H. G. and Hansma, P. K. (1989). Imaging Crystals, Polymers, and Processes in Water with the Atomic Force Microscope. *Science*, Vol. 243, No. 4898, pp. (1586-1589), 0036-8075

Ebeling, D., Holscher, H., Fuchs, H., Anczykowski, B. and Schwarz, U. D. (2006). Imaging of Biomaterials in Liquids: A Comparison between Conventional and Q-Controlled Amplitude Modulation ('Tapping Mode') Atomic Force Microscopy. *Nanotechnology*, Vol. 17, No. 7, pp. (S221-S226), 09574484

Gan, Y. (2009). Atomic and Subnanometer Resolution in Ambient Conditions by Atomic Force Microscopy. *Surface Science Reports*, Vol. 64, No. 3, pp. (99-121), 0167-5729

Gan, Y., Wanless, E. J. and Franks, G. V. (2007). Lattice-Resolution Imaging of the Sapphire (0001) Surface in Air by Afm. *Surface Science*, Vol. 601, No. 4, pp. (1064-1071)

Giessibl, F. J. (2000). Atomic Resolution on Si(111)-(77) by Noncontact Atomic Force Microscopy with a Force Sensor Based on a Quartz Tuning Fork. *Applied Physics Letters*, Vol. 76, No. 11, pp. (1470-1472), 00036951

Gould, S. A. C., Drake, B., Prater, C. B., Weisenhorn, A. L., Manne, S., Hansma, H. G., Hansma, P. K., Massie, J., Longmire, M. and et al. (1990). From Atoms to Integrated Circuit Chips, Blood Cells, and Bacteria with the Atomic Force Microscope. *Journal of Vacuum Science & Technology*, A, Vol. 8, No. 1, pp. (369-363), 0734-2101

Gupta, V., Hampton, M. A., Nguyen, A. V. and Miller, J. D. (2010). Crystal Lattice Imaging of the Silica and Alumina Faces of Kaolinite Using Atomic Force Microscopy. *Journal of Colloid and Interface Science*, Vol. 352, No. 1, pp. (75-80), 0021-9797

Gupta, V. and Miller, J. D. (2010). Surface Force Measurements at the Basal Planes of Ordered Kaolinite Particles. *Journal of Colloid and Interface Science*, Vol. 344, No. 2, pp. (362-371), 0021-9797

Hartman, H., Sposito, G., Yang, A., Manne, S., Gould, S. A. C. and Hansma, P. K. (1990). Molecular-Scale Imaging of Clay Mineral Surfaces with the Atomic Force Microscope. *Clays and Clay Minerals*, Vol. 38, No. 4, pp. (337-342)

Holscher, H., Ebeling, D. and Schwarz, U. D. (2006). Theory of Q-Controlled Dynamic Force Microscopy in Air. *Journal of Applied Physics*, Vol. 99, No. 8, pp. (84311-84311), 0021-8979

Israelachvili, J. N. (1985). *Intermolecular and Surface Forces: With Applications to Colloidal and Biological Systems* Academic Press

Johnsson, P. A., Eggleston, C. M. and Hochella, M. F. (1991). Imaging Molecular-Scale Structure and Microtopography of Hematite with the Atomic Force Microscope. *American Mineralogist*, Vol. 76, No. 7-8, pp. (1442-1445)

Klinov, D. and Magonov, S. (2004). True Molecular Resolution in Tapping-Mode Atomic Force Microscopy with High-Resolution Probes. *Applied Physics Letters*, Vol. 84, No. 14, pp. (2697-2699), 00036951

Kumai, K., Tsuchiya, K., Nakato, T., Sugahara, Y. and Kuroda, K. (1995). Afm Observation of Kaolinite Surface Using "Pressed" Powder. *Clay Science*, Vol. 9, No. 5, pp. (311-316), 0009-8574

Lindgreen, H., Garnaes, J., Besenbacher, F., Laegsgaard, E. and Stensgaard, I. (1992). Illite-Smectite from the North Sea Investigated by Scanning Tunnelling Microscopy. *Clay Minerals*, Vol. 27, No. 3, pp. (331-342)

Marcus, R. B., Ravi, T. S., Gmitter, T., Chin, K., Liu, D., Orvis, W. J., Ciarlo, D. R., Hunt, C. E. and Trujillo, J. (1989). Formation of Atomically Sharp Silicon Needles. 01631918, Washington, DC, USA, 1989

Marcus, R. B., Ravi, T. S., Gmitter, T., Chin, K., Liu, D., Orvis, W. J., Ciarlo, D. R., Hunt, C. E. and Trujillo, J. (1990). Formation of Silicon Tips with 1 Nm Radius. *Applied Physics Letters*, Vol. 56, No. 3, pp. (236-238), 0003-6951

Martin, Y., Williams, C. C. and Wickramasinghe, H. K. (1987). Atomic Force Microscope-Force Mapping and Profiling on a Sub 100-a Scale. *Journal of Applied Physics*, Vol. 61, No. 10, pp. (4723-4729), 0021-8979

Martinez, N. F., Patil, S., Lozano, J. R. and Garcia, R. (2006). Enhanced Compositional Sensitivity in Atomic Force Microscopy by the Excitation of the First Two Flexural Modes. *Applied Physics Letters*, Vol. 89, No. 15, pp. (153115-153111), 0003-6951

Masliyah, J. H. and Bhattacharjee, S. (2006). *Electrokinetic and Colloid Transport Phenomena* John Wiley & Sons, Inc.

Matsko, N. B. (2007). Atomic Force Microscopy Applied to Study Macromolecular Content of Embedded Biological Material. *Ultramicroscopy*, Vol. 107, No. 2-3, pp. (95-105), 0304-3991

Meyer, G. and Amer, N. M. (1990). Optical-Beam-Deflection Atomic Force Microscopy: The NaCl (001) Surface. *Applied Physics Letters*, Vol. 56, No. 21, pp. (2100-2101)

Moller, C., Allen, M., Elings, V., Engel, A. and Muller, D. J. (1999). Tapping-Mode Atomic Force Microscopy Produces Faithful High-Resolution Images of Protein Surfaces. *Biophysical Journal*, Vol. 77, No. 2, pp. (1150-1158), 0006-3495

Nagy, K. L. B. A. E. (1994). *Scanning Probe Microscopy of Clay Minerals* Clay Minerals Society, ISBN: 1881208087; 9781881208082 LCCN: 2005-282925, Boulder, CO

Ohnesorge, F. and Binnig, G. (1993). True Atomic Resolution by Atomic Force Microscopy through Repulsive and Attractive Forces. *Science (Washington, D. C., 1883-)*, Vol. 260, No. 5113, pp. (1451-1456), 0036-8075

Okajima, T., Sekiguchi, H., Arakawa, H. and Ikai, A. (2003). Self-Oscillation Technique for Afm in Liquids. *Applied Surface* Science, Vol. 210

Sharp, T. G., Oden, P. I. and Buseck, P. R. (1993). Lattice-Scale Imaging of Mica and Clay (001) Surfaces by Atomic Force Microscopy Using Net Attractive Forces. *Surface Science*, Vol. 284, No. 1-2, pp. (L405-L410)

Sokolov, I. Y. and Henderson, G. S. (2000). Atomic Resolution Imaging Using the Electric Double Layer Technique: Friction Vs. Height Contrast Mechanisms. *Applied Surface Science*, Vol. 157, No. 4, pp. (302-307), 0169-4332

Sokolov, I. Y., Henderson, G. S. and Wicks, F. J. (1999). Theoretical and Experimental Evidence for "True" Atomic Resolution under Non-Vacuum Conditions. *Journal of Applied Physics*, Vol. 86, No. 10, pp. (5537-5540)

Strus, M. C., Raman, A., Han, C. S. and Nguyen, C. V. (2005). Imaging Artefacts in Atomic Force Microscopy with Carbon Nanotube Tips. *Nanotechnology*, Vol. 16, No. 11, pp. (2482-2492), 0957-4484

Sugawara, Y., Ishizaka, T. and Morita, S. (1991). Scanning Force/Tunneling Microscopy of a Graphite Surface in Air. *Journal of Vacuum Science & Technology* , B, Vol. 9, No. 2, Pt. 2, pp. (1092-1095), 0734-211X

Vrdoljak, G. A., Henderson, G. S., Fawcett, J. J. and Wicks, F. J. (1994). Structural Relaxation of the Chlorite Surface Imaged by the Atomic Force Microscope. *American Mineralogist*, Vol. 79, No. 1-2, pp. (107-112)

Weihs, T. P., Nawaz, Z., Jarvis, S. P. and Pethica, J. B. (1991). Limits of Imaging Resolution for Atomic Force Microscopy of Molecules. *Applied Physics Letters*, Vol. 59, No. 27, pp. (3536-3538)

Wicks, F. J., Henderson, G. S. and Vrdoljak, G. A. (1994). Atomic and Molecular Scale Imaging of Layered and Other Mineral Structures. *CMS Workshop Lect.*, Vol. 7, No. Scanning Probe Microscopy of Clay Minerals, pp. (91-138)

Wicks, F. J., Kjoller, K., Eby, R. K., Hawthorne, F. C., Henderson, G. S. and Vrdoljak, G. A. (1993). Imaging the Internal Atomic Structure of Layer Silicates Using the Atomic Force Microscope. *Can Mineral*, Vol. 31, No. 3, pp. (541-550)

Wicks, F. J., Kjoller, K. and Henderson, G. S. (1992). Imaging the Hydroxyl Surface of Lizardite at Atomic Resolution with the Atomic Force Microscope. *Can. Mineral.*, Vol. 30, No. 1, pp. (83-91), 0008-4476

Wyckoff, R. W. G. (1968). *Crystal Structures* John Wiley & Sons, New York

4

Vibration Responses of Atomic Force Microscope Cantilevers

Thin-Lin Horng

Department of Mechanical Engineering, Kun-Shan University, Tainan
Taiwan, R.O.C.

1. Introduction

In this investigation, the solution of the vibration response of an atomic force microscope cantilever is obtained by using the Timoshenko beam theory and the modal superposition method. In dynamic mode atomic force microscopy (AFM), information about the sample surface is obtained by monitoring the vibration parameters (e.g., amplitude or phase) of an oscillating cantilever which interacts with the sample surface. The atomic force microscope (AFM) cantilever was developed for producing high-resolution images of surface structures of both conductive and insulating samples in both air and liquid environments (Takaharu et al., 2003 ; Kageshima et al., 2002 ; Kobayashi et al., 2002 ; Yaxin & Bharat, 2007). In addition, the AFM cantilever can be applied to nanolithography in micro/nano electromechanical systems (MEMS/NEMS) (Fang & Chang, 2003) and as a nanoindentation tester for evaluating mechanical properties (Miyahara, et al., 1999). Therefore, it is essential to preciously calculate the vibration response of AFM cantilever during the sampling process. In the last few years, there has been growing interest in the dynamic responses of the AFM cantilever. Horng (Horng, 2009) employed the modal superposition method to analyze the vibration responses of AFM cantilevers in tapping mode (TM) operated in a liquid and in air. Lin (Lin, 2005) derived the exact frequency shift of an AFM non-uniform probe with an elastically restrained root, subjected to van der Waals force, and proposed the analytical method to determine the frequency shift of an AFM V-shaped probe scanning the relative inclined surface in non-contact mode (Lin, et al., 2006). Girard et al. (Girard, et al., 2006) studied dynamic atomic force microscopy operation based on high flexure modes vibration of the cantilever. Ilic et al. (Ilic, et al., 2007) explored the dynamic AFM cantilever interaction with high frequency nanomechanical systems and determined the vibration amplitude of the NEMS cantilever at resonance. Chang et al. (Chang & Chu, 2003) found an analytical solution of flexural vibration responses on tapped AFM cantilevers, and obtained the resonance frequency at arbitrary dimensions and tip radii. Wu et al. (Wu, et al., 2004) demonstrated a closed-form expression for the sensitivity of vibration modes using the relationship between the resonant frequency and contact stiffness of the cantilever and the sample. Horng (Horng, 2009) developed an analytical solution to deal with the flexural vibration problem of AFM cantilever during a nanomachining process by using the modal superposition method.

The above studies considered the AFM cantilever as a Bernoulli-Euler beam model. The effects of transverse shear deformation and rotary inertia were assumed to be negligible in the analysis. However, for AFM-based cantilever direct mechanical nanomachining, the indentation and sampling of solid materials (e.g. polymer silicon and some metal surfaces) are performed. The effects of transverse shear deformation and rotary inertia in the vibration analysis should be taken into account for cantilevers whose cross-sectional dimensions are comparable to the lengths. Neglecting the effects of transverse shear deformation and rotary inertia in the vibration analysis may result in less accurate results. Hsu et al. (Hsu, et al., 2007) studied the modal frequencies of flexural vibration for an AFM cantilever using the Timoshenko beam theory and obtained a closed-form expression for the frequencies of vibration modes. However, the solution of the vibration response obtained using the modal superposition method for AFM cantilever modeled as a Timoshenko beam, and the response of flexural vibration of a rectangular AFM cantilever which has large shear deformation effects, are absent from the literature.

In this chapter, the response of flexural vibration of a rectangular AFM cantilever subjected to a sampling force is studied analytically by using the Timoshenko beam theory and the modal superposition method. Firstly, the governing equations of the Timoshenko beam model with coupled differential equations expressed in terms of the flexural displacement and the bending angle are uncoupled to produce the fourth order equation. Then, the sampling forces which are applied to the end region of the AFM cantilever by means of the tip, are transformed into an axial force, distributed transversal stress and bending stress. Finally, the response of the flexural vibration of a rectangular AFM cantilever subjected to a sampling force is solved using the modal superposition method. Moreover, a validity comparison for AFM cantilever modeling between the Timoshenko beam model and the Bernoulli-Euler beam model was conducted using the ratios of the Young's modulus to the shear modulus. From the results, the Bernoulli-Euler beam model is not suitable for AFM cantilever modeling, except when the ratios of the Young's modulus to the shear modulus are less than 1000. The Timoshenko beam model is a better choice for simulatimg the flexural vibration responses of AFM cantilever, especially for very small shear modulus.

2. Analysis

In contact mode, the AFM cantilever moves down by a small amplitude (1-5 nm) when the cantilever tip processes a sample surface. Therefore, the linear model can be used to describe the tip-sample interaction. The atomic force microscope cantilever, shown in Fig.1, is a small elastic beam with a length L, thickness H, width b, and a tip with a width of w and length h. x is the coordinate along the cantilever and $v(x,t)$ is the vertical deflection in the x-direction, as shown in Fig.2. One end of the cantilever, at $x = 0$, is clamped, while the other end, from $L-w$ to L, has a tip.

When the sampling is in progress, the tip makes contact with the specimen, resulting in a vertical reaction force, $F_y(t)$ and a horizontal reaction force, $F_x(t)$, both of which functions of time t. Assuming that the reaction forces act on the tip end, the product of the horizontal force and the tip length can form a bending stress on the bottom surface of the cantilever. The sampling system can be modeled as a flexural vibration motion of the cantilever. The motion is a function of mode shape and natural frequency, and its transverse displacement depends on time and the spatial coordinate x [7 and 8]. When the sampling forces are

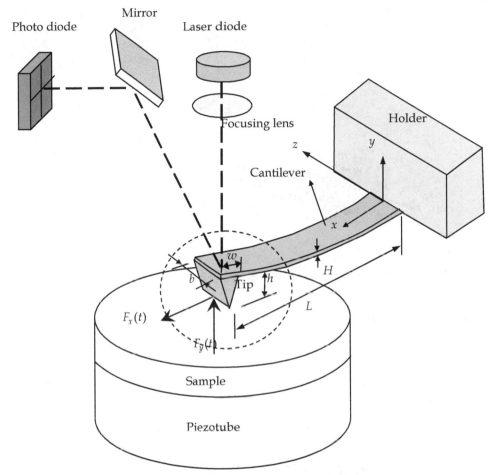

Fig. 1. Schematic diagram of an AFM tip-cantilever assembly processing a sample surface.

applied, the loads transmitted from the tip holder act on the end of the AFM cantilever, and can be modeled as the three parts shown in Fig.2, termed axial force $N(t)$, transverse excitation $p_l(x,t)$, and bending excitation $p_b(x,t)$.

Assuming that the transverse excitation is uniformly distributed on the bottom surface of the AFM cantilever, then it can be written as:

$$p_l(x,t) = F_y(t)u(x - L + w) / w,\tag{1}$$

where $u(x - L + w)$ is the unit step function.

The relationship between $F_x(t)$ and $F_y(t)$ can be expressed as $F_x = \dfrac{2\cos\theta}{\pi}F_y$ for a cone shape cantilever tip, where θ is the half-conic angle. The bending excitations, which result

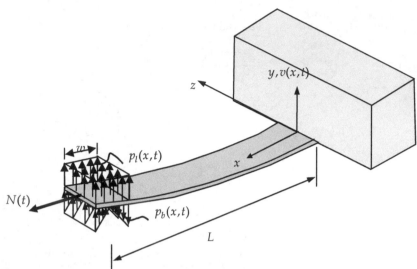

Fig. 2. Schematic diagram of excitations acting on the AFM cantilever

from the horizontal sampling force $F_x(t)$, act on the bottom surface of the AFM cantilever within the region from $L-w$ to L. They can be written as:

$$p_b(x,t) = \left[12h(2L-w-2x)\cos\theta / \pi w^3\right]F_y u(x-L+w) \tag{2}$$

By summing the above two excitations, the total transverse excitation $p_t(x,t)$ can be expressed as (Horng, 2009):

$$p_t(x,t) = C(x)F_y(t)u(x-L+w) \tag{3}$$

$$\text{where } C(x) = \frac{1+12h(2L-w)\cos\theta / \pi w^2}{w} - \frac{24h\cos\theta}{\pi w^3}x \tag{4}$$

Vibration behaviors of an AFM cantilever are examined using the Timoshenko beam theory. The effects of the rotary inertia and shear deformation are taken into account during contact with the sample. The governing equations of the Timoshenko beam model are two coupled differential equations expressed in terms of the flexural displacement and the angle of rotation due to bending. When the beam support is constrained to be fixed and all other external influences are set to zero, we obtain the classical coupled Timoshenko-beam partial differential equations (Hsu, et al., 2007):

$$\rho A \frac{\partial^2 v}{\partial t^2} - KAG(\frac{\partial^2 v}{\partial x^2} - \frac{\partial \psi}{\partial x}) = 0 \tag{5}$$

$$EI \frac{\partial^2 \psi}{\partial x^2} + KAG(\frac{\partial v}{\partial x} - \psi) - \rho I \frac{\partial^2 \psi}{\partial t^2} = 0 \tag{6}$$

where x is the distance along the center of the cantilever, $v(x,t)$ is the transverse displacement, t is time, $\psi(x,t)$ is the rotation of the neutral axis during bending, E is Young's modulus, G is shear modulus, I is the area moment of inertia, ρ is the volume density, K is the shear factor ($K = 5/6$ for rectangular cross-section), and A is the rectangular cross-sectional area of the cantilever.

Equations (5) and (6) may be uncoupled to produce a fourth order equation in $v(x,t)$. Considering the axial force effect, the classical uncoupled Timoshenko-beam partial differential equations can be written as (White, et al., 1995):

$$EI\frac{\partial^4 v(x,t)}{\partial x^4} + \rho A\frac{\partial^2 v(x,t)}{\partial t^2} - \frac{\partial}{\partial x}\left[N(t)\frac{\partial v(x,t)}{\partial x}\right] +$$

$$+\rho I\frac{\rho}{KG}\frac{\partial^4 v(x,t)}{\partial t^4} - \left(\rho I + \frac{\rho EI}{KG}\right)\frac{\partial^4 v(x,t)}{\partial x^2 \partial t^2} = p_t(x,t) \tag{7}$$

where $N(t) = F_x(t)$ is the axial force.

The mode-superposition analysis of a distributed-parameter system is equivalent to that of a discrete-coordinate system once the mode shapes and frequencies have been determined because in both cases, the amplitudes of the modal-response components are used as generalized coordinates in defining the response of the structure. In principle, an infinite number of these coordinates are available for a distributed-parameter system, since it has an infinite number of modes of vibration. Practically, however, only those modal components which provide significant contributions to the response need be considered (Ray & Joseph, 1993 ; William, 1998). The essential operation of the mode-superposition analysis is the transformation from the geometric displacement coordinates to the modal-amplitude or normal coordinates. For a one-dimensional system, this transformation is expressed as:

$$v(x,t) = \sum_{n=1}^{\infty} \varphi_n(x)Y_n(t) = \sum_{n=1}^{\infty} q_n(x,t) \tag{8}$$

where $q_n(x,t)$ is the response contribution of the n-th mode, $Y_n(t)$ is the normal coordinate, and $\varphi_n(x)$ is the n-th mode shape of the AFM cantilever. In order to find the natural frequencies and mode shapes, the following non-dimensional variables are defined:

$$\xi = \frac{x}{L}, \quad b^2 = \frac{\rho A L^4}{EI}\omega^2, \quad r^2 = \frac{I}{AL^2}, \quad s^2 = \frac{EI}{KAGL^2}. \tag{9}$$

Here ξ is the non-dimensional length along the beam, and ω is the radian frequency. Then, $\varphi_n(x)$ can be given by (White, et al., 1995):

$$\varphi_n(\xi) = C\left[\cosh b_n\alpha\xi - \frac{(R_1 - R_3)}{(R_2 - RR_4)}\sinh b_n\alpha\xi - \cos b_n\beta\xi + \frac{R(R_1 - R_3)}{(R_2 - RR_4)}\sin b_n\beta\xi\right] \tag{10}$$

where
$$\left\{ \begin{matrix} \alpha \\ \beta \end{matrix} \right\} = (1/\sqrt{2})\left\{ \mp (r^2 + s^2) + [(r^2 - s^2)^2 + 4/b_n^2]^{1/2} \right\}^{1/2} \tag{11}$$

$$R = \frac{(\alpha^2 + s^2)}{\alpha} \frac{\beta}{(\beta^2 - s^2)} \tag{12}$$

$$R_1 = (b_n / \alpha)\sinh b_n \alpha \tag{13}$$

$$R_2 = (b_n / \alpha)\cosh b_n \alpha \tag{14}$$

$$R_3 = (b_n / \beta)\sin b_n \beta \tag{15}$$

$$R_4 = -(b_n / \alpha)\cos b_n \beta \tag{16}$$

and b_n are the non-dimensional natural frequencies, which can be obtained using the characteristic equation

$$\left(\frac{\alpha^2 + s^2}{\alpha}\right)(R_3 R_4' - R_3' R_4 + R_4 R_1' - R_1 R_4') + \left(\frac{\beta^2 - s^2}{\beta}\right)(R_2 R_3' - R_2' R_3 + R_1 R_2' - R_2 R_1') = 0 \tag{17}$$

where
$$R_1' = [(\alpha^2 + s^2) / \alpha]b_n \alpha \cosh b_n \alpha \tag{18}$$

$$R_2' = [(\alpha^2 + s^2) / \alpha]b_n \alpha \sin b_n \alpha \tag{19}$$

$$R_3' = -[(\beta^2 - s^2) / \beta]b_n \beta \cos b_n \beta \tag{20}$$

$$R_4' = -[(\beta^2 - S^2) / \beta]b_n \beta \sin b_n \beta \tag{21}$$

Equation (8) simply states that any physically permissible displacement pattern can be modeled by superposing appropriate amplitudes of the vibration mode shapes for the structure. Substituting Eq. (8) into Eq. (7) and using orthogonally conditions gives

$$S_n \frac{d^4 Y_n(t)}{dt^4} + (M_n + T_n)\frac{d^2 Y_n(t)}{dt^2} + [-G_n F_x(t) + \omega_n^2 M_n]Y_n(t) = P_n(t) \tag{22}$$

where ω_n is the n-th mode natural frequency of the AFM cantilever obtained using:

$$\omega_n = b_n \sqrt{\frac{EI}{\rho A L^4}} \tag{23}$$

S_n, M_n, T_n and p_n are the generalized constants of the n-th mode, given by

$$S_n = (\rho I \frac{\rho}{KG})\int_0^L \varphi_n(x)^2 \, dx \tag{24}$$

$$M_n = (\rho A)\int_0^L \varphi_n(x)^2\, dx \tag{25}$$

$$T_n = (\rho I + \frac{\rho EI}{KG})\int_0^L \varphi_n(x)\frac{d^2\varphi_n(x)}{dx^2}\, dx \tag{26}$$

$$G_n(t) = \int_0^L \left[\varphi_n(x)\frac{d^2\varphi_n(x)}{dx^2}\right] dx \tag{27}$$

$$P_n(t) = \int_0^L \varphi_n(x)p_t(x,t)\, dx \tag{28}$$

Using Eq. (3) and Eq. (4), Eq. (28) can be rewritten as

$$P_n(t) = c_n F_y(t) \tag{29}$$

where

$$c_n = \int_0^L \varphi_n(x)\left(\frac{1+12h(2L-w)\cos\theta\,/\,\pi w^2}{w} - \frac{24h\cos\theta}{\pi w^3}x\right)u(x-L+w)\,dx \tag{30}$$

Then, the Normal-Coordinate Response Equation, which is exactly the same equation considered for the discrete-parameter case, can be solved.

$$\frac{d^4Y_n(t)}{dt^4} + (M_n + T_n)/S_n\frac{d^2Y_n(t)}{dt^2} + \frac{[-G_n F_x(t) + \omega_n^2 M_n]}{S_n}Y_n(t) = c_n F_y(t)/S_n \tag{31}$$

Assuming a zero initial condition, with $v(x,0)=0$, $\dot{v}(x,0)=0$, $\ddot{v}(x,0)=0$ and $\dddot{v}(x,0)=0$, and providing that the sampling force $F_y(t)$ is a series of harmonics, $F_y(t)$ can be written as:

$$F_y(t) = \sum_{i=1}^m F_i \sin(\omega_i t) \tag{32}$$

When the j-th excitation frequency ω_j is equal to the n-th natural frequency ω_n, the Runge-Kutta method is introduced to solve the above fourth-order system.

3. Results and discussion

The main goal of this study is to analyze the flexural vibration responses in nanoscale processing using atomic force microscopy modeled as a Timoshenko beam. To demonstrate the validity of the analytical solution, numerical computations were performed. The geometric and material parameters considered were as follows:

$E = 170Gpa$, $\bar{m} = 0.2898km\,/\,\mu m$, $\rho = 2300km\,/\,m^3$, $L = 125\mu m$, $b = 30\mu m$, $H = 4.2\mu m$, $h = 5\mu m$, $2\theta = 30°$ and $m = 3$, $F_i = 1000\,/\,(2i-1)\times 10^{-9}$.

The modulus-ratio REG, defined as the ratio of E to G (i.e. $REG = E/G$), is introduced to define the values of shear modulus G and to describe the effects of shear deformation. In this study, the flexural vibration responses at the end of the AFM cantilever were obtained using the contribution of the first five vibration modes. A non-dimensional response was used to normalize the static response as given in $F_1L^3 / (3EI)$, and $\omega_i = (2i-1) \times r\omega_n$, and set as the simulated values of the excitation frequency of the vertical sampling force. Thus, $F_y(t)$ is taken as:

$$F_y(t) = 1000\left(\sin r\omega_n t + \frac{1}{3}\sin 3 \times r\omega_n t + \frac{1}{5}\sin 5 \times r\omega_n t \right)(nN) \tag{33}$$

where r is the frequency ratio that can used to describe the deviation between the excitation frequency and modal frequencies. The modal frequency ω_n and modal shapes $\varphi(x)$ of the first five vibration modes for an AFM cantilever are shown in Fig.3 and Fig.4, respectively.

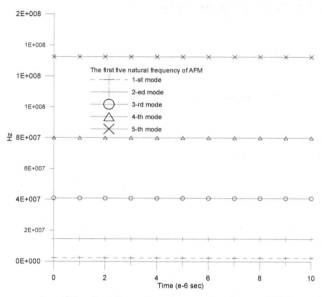

Fig. 3. Natural frequencies of the first five vibration modes for an AFM cantilever.

In order to investigate the effects of transverse shear deformation, the response histories at the end point of the AFM cantilever between different small and large modulus-ratios, with respect to excitation frequencies far away from($r = 0.1$) and close to ($r = 0.9$) the first natural frequency, are shown in Fig.5 to Fig.8. Figure 5 and Fig.6 indicate that the responses are similar for the various modulus-ratios when they are less than 1000. This means that if the effects of transverse shear deformation are small enough to be negligible, the Timoshenko beam model can be reduced to the Bernoulli-Euler beam model. Figure 6 also reveals that the resonance effect occurs when the AFM cantilever has small modulus-ratios and the excitation frequencies are close to the modal frequencies.

Figure 7 and Fig.8 show the response histories at the end point of the AFM cantilever between different large modulus-ratios for excitation frequencies far away from ($r = 0.1$) and close to ($r = 0.9$) the first modal frequency, respectively. The results are quite different

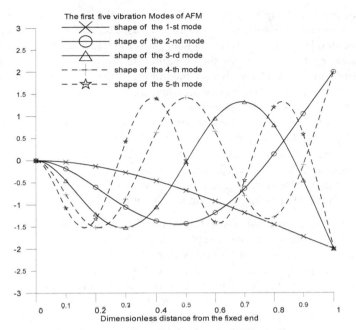

Fig. 4. The shape of the first five vibration modes for an AFM cantilever.

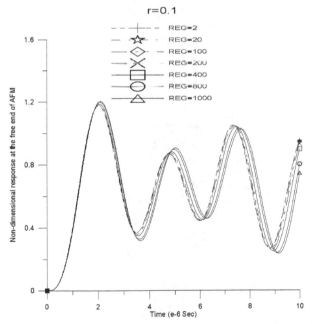

Fig. 5. The effect of various small modulus-ratios (REG) on the response of the end point for the excitation frequency far away from the first natural frequency, i.e. $0.1\omega_1$.

from those of Fig.5 and Fig.6. Figure 7 indicates that the magnitude of the transversal response increases and its oscillating frequency decreases when the modulus-ratio increases. This is because that large shear deformation increases the transversal response, which slows down the oscillating frequency when the excitation frequency is far away from the natural frequency. However, Fig. 8 tells us that the magnitude of the transversal response decreases and its oscillating frequency becomes small when the modulus-ratio increases. The reason for this is that the effects of resonance were counteracted by the transverse shear deformation, resulting in the small transversal response when the AFM cantilever has the sufficiently large modulus-ratios and the excitation frequencies of AFM cantilever are close to the modal frequencies. Consequently, Fig.7 and Fig.8 imply that when a sufficiently small shear modulus is used in AFM cantilever, the effect of transverse shear deformation has a significant effect on the transversal response and the Timoshenko beam model is the proper choice for simulating AFM cantilever dynamic behavior.

Figures 9 shows the effects of various tip holder widths, w, on the response of the end point. The widths are normalized by the length of the AFM cantilever. Fig.9 shows that the response at the free end decreases as the width of the tip increases. Therefore, an AFM cantilever with a large tip width is suggested to reduce the response at the end of the AFM cantilever. Figure 10 shows the response histories at the end point of the AFM tip for various tip lengths h. From the simulation results shown in Fig.10, the responses are relatively small when the tip length is large. This is due to a large tip length producing large bending effects. Therefore, an AFM tip with large tip length is suggested to reduce the response at the end of the AFM cantilever.

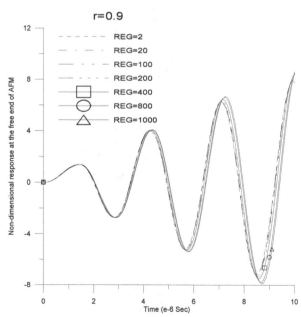

Fig. 6. The effect of various small modulus-ratios (REG) on the response of the end point for the excitation frequency close to the first natural frequency, i.e. $0.9\omega_1$.

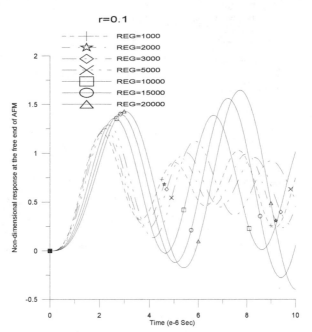

Fig. 7. The effect of various large modulus-ratio (REG) on the response of the end point for the excitation frequency far away from the first natural frequency, i.e. $0.1\omega_1$

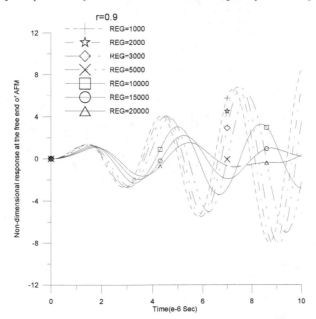

Fig. 8. The effect of various large modulus-ratio (REG) on the response of the end point for the excitation frequency which is close to the first natural frequency, i.e. $0.9\omega_1$.

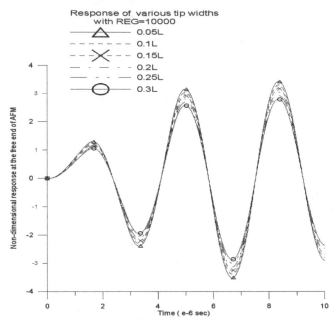

Fig. 9. Response histories at the end point for various tip widths with $REG = 10000$.

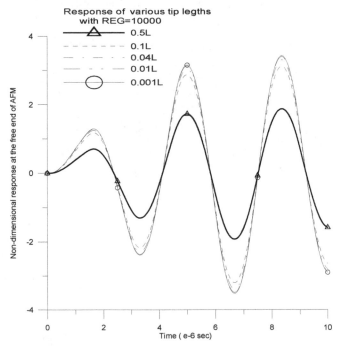

Fig. 10. Response histories at the end point for various tip lengths with $REG = 10000$.

4. Conclusions

The modal superposition method and the Timoshenko beam theory were applied to determine the flexural vibration responses at the end of the AFM cantilever during AFM-based nanoprocessing process. As expected, the Bernoulli-Euler beam model for AFM cantilever applies to the small effects of transverse shear deformation, but not for modulus-ratios greater than 1000. When modulus-ratios are greater than 1000, the Timoshenko beam model is the proper choice for simulating the flexural vibration responses of AFM cantilever. Moreover, the oscillating frequency of transversal response decreases due to the transverse shear deformation and the magnitudes of the transversal response depend on the deviation between the excitation frequencies and the modal frequencies. In conclusion, one can reduce the response at the end of AFM cantilever by decreasing the shear modulus when the frequencies of processing are far away from the modal frequencies, and by increasing the shear modulus when the frequencies of processing are close to the modal frequencies. Furthermore, an AFM cantilever with a large tip width and length is suitable for reducing the response at the end of the AFM cantilever.

5. Acknowledgements

This work was supported by the National Science Council, Taiwan, Republic of China, under grant NSC 99-2221-E-168-021.

6. References

Chang, W.J. & Chu, S.S. (2003). Analytical solution of flexural vibration response on taped atomic force microscope cantilevers, Phys. Letter A. Vol. 309, pp. 133-137.

Fang, T.H. & Chang, W.J. (2003). Effects of AFM-based nanomachining process on aluminum surface, J. Phys. Chem. Solids, *J. Phys. Chem. Solids*, Vol. 64 913-918.

Girard, P.; Ramonda, M. & R. Arinero, (2006). Dynamic atomic force microscopy operation based on high flexure modes of the cantilever, *Rev. Sci. Instrum.* Vol. 77, 096105.

Horng, T.L. (2009). Analytical Solution of Flexural Vibration Responses on Nanoscale Processing Using Atomic Force Microscopy, *J. Mater. Pro. Tech.*, Vol. 209, pp. 2940-2945.

Horng, T.L. (2009). Analyses of Vibration Responses on Nanoscale Processing in a Liquid Using Tapping-Mode Atomic Force Microscopy, *Appl. Surf. Sci.* Vol. 256 311-317.

Hsu, J.C.; Lee, H.L.& Chang, W.J. (2007). Flexural Vibration Frequency of Atomic Force Microscope Cantilevers Using the Timoshenko Beam Model, *Nanotechnology.* Vol. 18, 285503.

Ilic, B.; Krylov, S.; Bellan L.M. & H.G. Craighead, (2007). Dynamic characterization of nanoelectromechanical oscillators by atomic force microscopy, J. *Appl. Phys.* Vol. 101, 044308

Kageshima, M.; Jensenius, H.; Dienwiebel, M.; Nakayama, Y.; Tokumoto, H. ; Jarvis, S.P. & Oosterkamp, T.H. (2002). Noncontact atomic force microscopy in liquid environment with quartz tuning fork and carbon nanotube probe. *Appl. Surf. Sci.* Vol. 188, pp.440-444.

Kobayashi, K.; Yamada, H. & Matsushige, K. (2002). Dynamic force microscopy using FM detection in various environments. Appl. Surf. Sci. Vol.188, pp. 430-434.

Lin, S.M. (2005). Exact Solution of the frequency shift in dynamic force microscopy, *Appl. Surf. Sci.* Vol. 250, pp. 228-237.

Lin, S.M.; Lee. S.Y. & B-S Chen, (2006). Closed-form solutions for the frequency shift of V-shaped probes scanning an inclined surface, *Appl. Surf. Sci.* Vol. 252 6249-6259.

Miyahara, K.; Nagashima, N.; Ohmura T. & Matsuoka, S. (1999). Evaluation of mechanical properties in nanometer scale using AFM-based nanoindentation tester, *Nanostruct. Mater.* Vol. 12, pp.1049-1052.

Ray, W. & Joseph, P. (1993). Dynamic of Structure, second ed., McGraw-Hill, Inc., New Jersey.

Takaharu Okajima; Hiroshi Sekiguchi; Hideo Arakawa & Atsushi Ikai. (2003). Self-oscillation technique for AFM in liquids. *Appl. Surf. Sci.* Vol. 210, pp.68-72.

White, M. W. D. & Heppler, G. R. (1995). Vibration Modes and Frequencies of Timoshenko Beams with Attached Rigid Bodies, *ASME J. Appl. Mech.* Vol. 62, pp.193-199.

William, T. (1998). Theory of Vibrations with Application, fifth ed., Prentice Hall, New Jersey.

Wu, T.S.; Chang, W.J. & J.C. Hsu, (2004). Effect of tip length and normal and lateral contact stiffness on the flexural vibration response of atomic force microscope cantilevers, *Microelectronic Eng.* Vol. 71, pp.15-20.

Yaxin Song & Bharat Bhushan, (2007). Finite-element vibration analysis of tapping-mode atomic force microscopy in liquid, *Ultramicroscopy.* Vol. 107, pp. 1095-1104.

5

Wavelet Transforms in Dynamic Atomic Force Spectroscopy

Giovanna Malegori and Gabriele Ferrini

Interdisciplinary Laboratories for Advanced Materials Physics (i-LAMP) and Dipartimento di Matematica e Fisica, Università Cattolica del Sacro Cuore, I-25121 Brescia Italy

1. Introduction

Since their invention scanning tunneling microscopy (STM, (Binnig et al., 1982)) and atomic force microscopy (AFM, (Binnig et al., 1986)) have emerged as powerful and versatile techniques for atomic and nanometer-scale imaging. In this review we will focus on AFM, whose methods have found applications for imaging, metrology and manipulation at the nanometer level of a wide variety of surfaces, including biological ones (Braga & Ricci, 2004; Garcia, 2010; Jandt, 2001; Jena & Hörber, 2002; Kopniczky, 2003; Morita et al., 2009; 2002; Yacoot & Koenders, 2008). Today AFM is regarded as an essential tool for nanotechnology and a basic tool for material science in general.

AFM relies on detecting the interaction force between the sample surface and the apex of a sharp tip protruding from a cantilever, measuring the cantilever elastic deformation (usually its bending or twisting) caused by the interaction forces. Fig. 1a shows a schematic interaction force dependence on tip-sample distance in vacuum (Hölscher et al., 1999). As the distance between the cantilever and the sample surface is reduced by means of a piezoelectric actuator, the tip first experiences an attractive (typically van der Waals) force, that increases to a maximum value. During further approach, the attractive force is reduced until a repulsive force regime is reached. Therefore the AFM is a sensitive force gauge on the nanometer and atomic scale (Butt et al., 2005; Cappella & Dietler, 1999; Garcia & Perez, 2002; Giessibl, 2003; Mironov, 2004).

The use of AFM in such tip-sample force measurements is commonly referred to as *force spectroscopy*. The simplest technique used for quantitative force measurements involves directly monitoring the static deflection of the cantilever as the tip moves towards the surface (approach curve) and then away (retraction curve), providing a deflection versus distance plot. To obtain a force-distance curve, the cantilever deflection is converted to tip-sample interaction force using Hooke's law (Butt et al., 2005; Cappella & Dietler, 1999), after calibration of the cantilever spring constant (Hutter & Bechhoefer, 1993; Sader, 1999).

A typical force curve at room temperature and in air is shown in Fig. 1b (Butt et al., 2005; Cappella & Dietler, 1999). During the approach to the surface, an attractive long-range force on the probe bends the cantilever toward the surface. Then the tip suddenly jumps into contact with the surface due to the large gradient of the attractive force near the sample surface

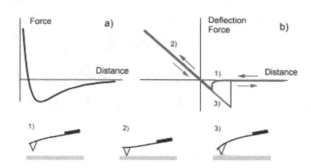

Fig. 1. a) Schematic diagram of the tip-sample forces as a function of distance in vacuum. b) Schematic picture of a typical force curve in air, showing the cantilever deflection (proportional to the applied force) versus the tip-sample distance during approach (red) and withdrawing (blue). The cantilever interactions in the various zones of the force curve are illustrated below: 1) attractive long-range interaction, 2) repulsive contact interaction, 3) adhesive capillary force.

(van der Waals, adhesion, capillary forces, electrical interactions). A further approach of the cantilever to the sample leads to an increasing cantilever deflection in the opposite direction due to repulsive contact force. Finally, during the retraction curve, the tip-sample separation (jump-off contact) occurs at distance larger than the jump to contact distance. The hysteresis is usually due to adhesive capillary force which keeps the tip in contact to the sample.

Depending on their compliance, the cantilever, tip and sample may experience an elastic or plastic deformation. In case of elastic interaction, the linear parts of the force-distance curve recorded during forward and reverse motion coincide. For compliant samples, such as biological samples, the shape of the curves is more complex due to indentation effects. Generally, the force-distance curve provides information on the nanoscale tip-surface interactions, the local elasticity of the sample and the adhesion forces (Andreeva et al., 2010; Butt et al., 2005; Cappella & Dietler, 1999; Radmacher et al., 1996).

More refined techniques, suitable for atomic scale investigations in vacuum and in liquids, rely on measuring the dynamic parameters of the cantilever excited at or near its resonant frequency while experiencing the force field of the sample surface. The interactions of the tip with the sample surface perturbs the amplitude, frequency, phase or damping of the cantilever oscillation. The tip-sample interaction force can be inferred by the modifications of these parameters (Albrecht et al., 1991; Crittenden et al., 2005; Hölscher et al., 1999; Lantz et al., 2001; Martin et al., 1987; Palacios-Lidón & Colchero, 2006; Sugimoto, Innami, Abe, Custance & Morita, 2007).

There are two basic methods in dynamic AFM operations, the Amplitude Modulation (AM) technique (Martin et al., 1987) and the Frequency Modulation (FM) technique (Albrecht et al., 1991).

AM-AFM detect the variation of amplitude and phase of the excited cantilever oscillations due to tip-sample interactions and has been successfully implemented in ambient conditions involving repulsive tip-sample interactions, the so called "tapping mode" . Though extremely interesting, these techniques will not be considered further in this review and the interested reader is referred to the bibliography (Garcia, 2010; Garcia & Perez, 2002).

FM-AFM detects with high sensitivity minute changes in the cantilever resonant frequency under a particular feedback mode, while the tip approaches the surface (Giessibl, 2003; Morita et al., 2009; 2002). The tip-sample force versus distance can be inferred by inverting the resonance frequency shift versus distance curves (Giessibl, 2003; Hölscher et al., 1999; Sader & Jarvis, 2004). This technique detects long-range electrostatic and van der Waals forces as well as repulsive short-range force providing the chemical identification of single atoms at surfaces. In fact, short-range interaction depends primarily on the chemical identity of the atoms involved since they are associated with the onset of the chemical bond between the outermost atom of the tip apex and the surface atoms. Then precise measurement of such short-range chemical force allows to distinguish between different atomic species even though they exhibit very similar chemical properties and identical surface position preferences so that any discrimination attempt based on topographic measurements is impossible (Gross et al., 2009; Lantz et al., 2001; Sugimoto, Pou, Abe, Jelinek, Pérez, Morita & Custance, 2007). Moreover, FM-AFM has recently attained atomic scale resolution in liquids, (Fukuma, 2010; Fukuma et al., 2010).

Three dimensional frequency shift maps over a surface have been acquired too, a method known as 3D AFM spectroscopy. Measuring the frequency shift and the topography over finely spaced planes parallel to the sample surface, allows to apply drift corrections to the data and retrieve a three dimensional frequency shift grid. From these data, the interaction forces, by inverting each frequency-shift versus distance curve, the tip-sample potential energy and the energy dissipated per oscillatory cycle are obtained (Albers et al., 2009).

An interesting alternative to performing force spectroscopy is the broad band excitation (BE) scheme which takes advantage of the fact that the simultaneous analysis at all frequencies within the excited band reduces the acquisition time. Broad band excitation can be achieved by an external driving force (Jesse et al., 2007) or by thermal excitation (Malegori & Ferrini, 2010a; Roters & Johannsmann, 1996; Vairac et al., 2003).

In the first case, instead of a simple sinusoidal excitation, the BE method uses a driving signal with a predefined spectral content in the frequency band of interest. The cantilever response to the BE drive is measured and Fourier transformed to yield the amplitude and phase-frequency curves at several distances from the surface. The measured response curves allow to calculate the resonant frequency, amplitude, and Q-factor at each distance and display the data as a 2D image. Again, the force versus distance curve can be inferred from the evolution of the frequency, the energy dissipation from the phase and quality factor modifications. The fast Fourier transform/fitting routine replaces the traditional lock-in/low pass filter that provides amplitude and phase at a single frequency at time. In the BE method the system is excited and the response is measured simultaneously at all frequencies within the excited band (parallel excitation/detection). On the contrary, standard lock-in detectors acquire the response over a broad band by sampling one frequency at time. In both cases the complete spectral acquisition is carried out at several tip-sample separation. The BE approach allows a significant reduction of the acquisition time by performing the detection on all frequencies in parallel, so that an extremely broad frequency range (25-250 kHz) can be probed in ~1 s. A comparable scan over the same frequency band using a lock-in would require ~30 min.

Another possibility in broad band excitation is the *thermal excitation*, essentially a random force of thermal origin due to the interaction of the cantilever with the surrounding environment (the thermal reservoir). The random force power spectrum does not depend on frequency and produces the so called Brownian motion or white noise (Callen & Greene, 1952). A fundamental point is that the Brownian motion of the thermally driven cantilever is modified by the tip-sample interaction forces. It follows that the temporal trace of the cantilever thermal fluctuation detected at various distances from the surface contains informations on the tip-sample interaction and can be analyzed to reconstruct the tip-sample potential and interaction force.

Three different approaches are possible: (a) to measure the shift of the cantilever resonant frequency of the first flexural modes as the tip moves toward the surface to retrieve the gradient of the interaction forces (Roters & Johannsmann, 1996); (b) to detect the mean square displacement of the tip subjected to thermal motion in order to estimate the interaction force gradient dependence on the tip-sample distance (Malegori & Ferrini, 2010a); (c) to analyze the probability distribution of the tip's position during the Brownian motion. Then the Boltzmann distribution is used to calculate the Helmholtz free energy of the tip interacting with the surface as a function of the tip-sample distance. The interaction force gradient is inferred from the second derivative of the Helmholtz free energy (Cleveland et al., 1995; Heinz et al., 2000; Koralek et al., 2000). In (Malegori & Ferrini, 2010a) the three methods have been applied simultaneously to the same experimental session to compare their peculiarities.

BE methods provides the lower limits on the acquisition time required to detect a complete force versus distance curve (0.1-1 s). Nevertheless, this acquisition time is still too long and incompatible with the rate of 1-30 ms/pixel which is the value necessary to obtain a complete force image. This motivated the introduction of the wavelet transform in thermally excited dynamic spectroscopy, a new approach to spectroscopy measurements that is the topic of this review. Wavelet transforms allow to reduce the acquisition time to values compatible with practical dynamic force spectroscopy imaging and to apply the analysis simultaneously to all the cantilever modes, either flexural and torsional, within the cut-off frequency of the acquisition system.

2. Thermally excited cantilever: the Brownian motion

2.1 Fluctuation-dissipation theorem

At non zero absolute temperature, a system in thermodynamic equilibrium is not at rest but continuously fluctuates around its equilibrium state. For example, a mechanical system in equilibrium at temperature T, continuously exchanges its mechanical energy with the thermal energy of the thermal bath in which it is immersed.

To analyze the thermal fluctuations of a system, consider an extensive variable x which is coupled to the intensive parameter F in the Hamiltonian of the system (Callen & Greene, 1952; Paolino & Bellon, 2009). In the frequency domain, the response function $G(\omega)$ describes the response of the system to the driving variable F. It is defined as:

$$G(\omega) = \frac{F(\omega)}{x(\omega)}$$

The thermal fluctuations of the observable x are described by the Fluctuation-Dissipation theorem (Gillespie, 1993; 1996) which connects the power spectral density (PSD) of the fluctuations of the variable x, $S_x(\omega)$, to the temperature and the response function $G(\omega)$ as:

$$S_x(\omega) = \frac{x^2(\omega)}{\Delta\omega} = -\frac{2k_B T}{\omega}\Im\left[\frac{1}{G(\omega)}\right] \tag{1}$$

Here k_B, $\Delta\omega$ and $\omega = 2\pi f$ are the Boltzmann constant, the angular frequency bandwidth and the angular frequency or pulsation associated to frequency f and $\Im$ denotes the imaginary part. The average shape of a spontaneous fluctuation is identical with the observed shape of a macroscopic irreversible decay toward equilibrium and is, therefore, describable in terms of the macroscopic response function.

2.2 Dynamic response of the cantilever

An example of mechanical dissipative system is the AFM cantilever placed in air far from the surface (free cantilever) and driven by background thermal energy. The cantilever is in thermal equilibrium with the molecules of the fluid in which it is immersed. In this situation it fluctuates mainly in response to stochastic forces due to the molecular motion from the temperature of the thermal bath.

The cantilever is described as an elastic massless beam (with elastic constant k). One end is fixed to the chip whereas a mass m (the tip) is localized on the free end. Then the cantilever dynamics can be reasonably modeled as a stochastic harmonic oscillator with viscous dissipation (Gillespie, 1996; Paolino & Bellon, 2009; Shusteff et al., 2006). In this case x is the displacement from equilibrium of the tip and F is the force applied to the system. For a non-interacting cantilever, the external driving force F is the thermal stochastic force F_{th}, which accounts for the interaction with the local environment. The resulting Brownian motion of the tip displacement x is described by the second order Langevin equation

$$m\ddot{x}(t) + \gamma\dot{x}(t) + kx(t) = \Gamma_{th}(t) \tag{2}$$

where γ is the damping factor.

F_{th} is defined by its statistic properties (Langevin hypothesis (Gillespie, 1993; 1996)) as a zero-mean force ($\langle F_{th}(t)\rangle = 0$), completely uncorrelated in time

$$\langle F_{th}(t)F_{th}(t+\tau)\rangle = \alpha\delta(\tau) \tag{3}$$

Here the brackets $\langle\rangle$ denote time averaging, $\delta(\tau)$ is the Dirac delta function and α a proportionality constant which will be determined through the fluctuation-dissipation theorem.

The Wiener-Khintchine theorem (Callen & Greene, 1952; Gillespie, 1996) states that, for a stationary random process, the power spectral density and its temporal autocorrelation function are mutual Fourier transforms. Applying the theorem to the the power spectrum of the thermal activating force $S_F(\omega)$ we obtain

$$S_F(\omega) = \hat{A}_F(\omega) = \alpha \tag{4}$$

where $\hat{A}_F(\omega)$ is the Fourier transform of the autocorrelation function $A_F(\tau) = \langle F_{th}(t)F_{th}(t+\tau)\rangle$. The stochastic force of the physical system is called *white noise* because it has no frequency dependence.

Now it is possible to connect the correlation function, that characterizes the fluctuating forces, to the dissipative part of the equation of motion, i.e. the damping factor γ.

The transfer function of the system is provided by writing Eq. 2 in Fourier space (frequency domain) as

$$G(\omega) = k\left[1 - \frac{\omega^2}{\omega_0^2} + i\frac{\omega}{Q\omega_0}\right] \tag{5}$$

where we introduced the resonant angular pulsation $\omega_0 = \sqrt{k/m}$ and the quality factor $Q = m\omega_0/\gamma$. The PSD of the thermal fluctuations x, using Eq. 1, is given by

$$S_x(\omega) = \frac{2k_BT}{k\omega_0}\frac{1/Q}{(1 - \omega^2/\omega_0^2)^2 + (\omega/\omega_0 Q)^2} \tag{6}$$

which is a Lorentzian curve with linewidth given by $\Delta\omega = \omega_0/Q$.

The PSD of the fluctuations $S_x(\omega)$ is related to the power spectrum of the stochastic thermal activating force $S_F(\omega)$ through the response function $G(\omega)$ (Shusteff et al., 2006) by $S_F(\omega) = S_x(\omega)G^2(\omega)$. Then from Eq. 5 and Eq. 6 we obtain

$$S_F(\omega) = 2k_BT\gamma \tag{7}$$

The constant α is determined by Eqs. 4 and 7 as $\alpha = 2k_BT\gamma$, providing an autocorrelation function of the external stochastic force expressed by:

$$\langle F_{th}(t)F_{th}(t+\tau)\rangle = 2k_BT\gamma\delta(\tau)$$

The last relation is another expression of the Fluctuation-Dissipation theorem which quantifies the intimate connection between the viscous coefficient γ and the randomly fluctuating force $F_{th}(t)$. It implies that the stochastic fluctuating force is an increasing function of γ and vanishes if and only if γ vanishes. The dissipative damping force $-\gamma\dot{x}$ and the fluctuating force F_{th} are correlated so that one cannot be present without the other one. This is because the microscopic events that give rise to those two forces simply cannot be separated into one kind of microscopic event (like the molecular collision) that gives rise *only* to a viscous effect and another kind that gives rise *only* to a fluctuating effect.

The Parseval relation is used to determine the variance of the fluctuations of the observable x (in our case the cantilever positional fluctuations) by integrating the positional PSD $S_x(\omega)$. Then:

$$\langle x^2\rangle = \int_{-\infty}^{+\infty} S_x(\omega)d\omega = \frac{2k_BT}{k\omega_0 Q}\int_{-\infty}^{+\infty} \frac{d\omega}{\left(1 - \frac{\omega^2}{\omega_0^2}\right)^2 + \left(\frac{\omega}{Q\omega_0}\right)^2} = \frac{k_BT}{k} \tag{8}$$

The potential energy of the cantilever modeled as a damped harmonic oscillator takes the form of:

$$\frac{1}{2}m\omega_0^2\langle x^2\rangle = \frac{1}{2}k\langle x^2\rangle = \frac{1}{2}k_BT$$

where $\langle x^2 \rangle$ represents the mean square displacement of the cantilever caused by the thermal motion in the direction normal to the surface. This relation is an expression of the equipartition theorem, stating that in a system in thermal equilibrium every independent quadratic term in its total energy has a mean value equal to $1/2k_B T$. An analysis of the cantilever thermal motion which explicitly considers all possible vibration modes can be found in (Butt & Jaschke, 1995).

Finally, we would like to point out that the reverse path is also possible, by demonstrating the Fluctuation-Dissipation theorem from the equipartition theorem, see (Gillespie, 1993; Shusteff et al., 2006).

2.3 Cantilever in interaction

Near the surface, the tip experiences an interaction force $F_{ts}(z)$, which depends on the distance $z = z(t)$ between the probe apex and the surface. The force is positive along the surface normal direction. The cantilever motion is now described by

$$m\ddot{x}(t) + \gamma\dot{x}(t) + kx(t) = F_{th}(t) + F_{ts}(z) \tag{9}$$

where x is the displacement from the equilibrium position of the free cantilever, see Fig. 2.

Fig. 2. Schematic representation of the variables describing the cantilever motion. z is the instantaneous tip-sample distance, positive along the surface normal direction. x is the instantaneous displacement from the equilibrium position of the free cantilever, negative when the cantilever is bent toward the sample. z_0 is the average tip-sample distance and x_0 the corresponding average tip displacement from the equilibrium under static interaction forces. $x' = x - x_0 = z - z_0$ is the cantilever displacement from the average equilibrium position under static interaction.

For small oscillations of x and z around the equilibrium position of the cantilever under static interaction, indicated as x_0 and z_0, the derivative of the force can be considered constant for the whole range covered by the oscillating cantilever. Therefore the force may be approximated by the first (linear) term in the series expansion (Giessibl, 2003; Mironov, 2004):

$$F_{ts}(z) = F_{ts}(z_0) + \frac{\partial F_{ts}}{\partial z}(z_0)(z - z_0)$$

The constant term of the force $F_{ts}(z_0)$ statically deflects the cantilever in the new equilibrium position $x_0 = F_{ts}(z_0)/k$. The interaction force gradient influences the cantilever oscillations around this position. By introducing the displacement from the equilibrium position under static interaction $x' = x - x_0 = z - z_0$, which incorporates the cantilever static bending, (see

Fig. 2), we come to the equation

$$m\ddot{x}'(t) + \gamma\dot{x}'(t) + \left(k - \frac{\partial F_{ts}}{\partial z}(z_0)\right)x'(t) = F_{th}(t) \tag{10}$$

This means that in case of small oscillations, as for instance the thermally excited oscillations, the presence of a force gradient results in a change of effective stiffness of the system

$$k^* = k - \frac{\partial F_{ts}}{\partial z} \tag{11}$$

Since $\omega_0' = 2\pi f_0' = \sqrt{k^*/m}$, the cantilever resonance frequency is modified as

$$f_0' = f_0\sqrt{1 - \frac{1}{k}\frac{\partial F_{ts}}{\partial z}}$$

In case of small force gradient, $|\partial F_{ts}/\partial z| << k$, the shift in eigenfrequency $\Delta f = f_0' - f_0$ becomes proportional to the force gradient

$$\frac{\partial F_{ts}}{\partial z} = -2k\frac{\Delta f}{f_0} \tag{12}$$

Therefore, one can determine the tip-sample force gradient by measuring the frequency shift Δf. Approaching the surface, the attractive tip-sample force causes a sudden jump-to-contact. In the quasi-static mode, the instability occurs at a distance z_{jtc} where

$$\left|\frac{\partial F_{ts}}{\partial z}(z_{jtc})\right| > k \tag{13}$$

so that only the long range part of the interaction curve is accessible (Giessibl, 2003; Hutter & Bechhoefer, 1993). The jump-to-contact effect can be avoided by using stiff cantilevers and dynamic methods such as FM-AFM.

3. Time meets frequency, the mathematical framework

3.1 Fourier transform

Experimental data in dynamic atomic force spectroscopy frequently appear as a time series. Time series often are transformed in the frequency domain to describe their spectral content. A typical method for signal processing is the Fourier transform (FT). As a paradigmatic example, we will describe how Fourier analysis can be used to analyze the temporal trace of the cantilever thermal vibrations, detected by a standard AFM optical beam deflection system, and the kind of information possibly extracted. Finally, the limitations of this approach will be discussed.

The power spectral density (PSD) spectrum of the time signal, extending over a temporal interval sufficiently long to assure the needed spectral resolution, reveals peaked structures corresponding to the various oscillation eigenmodes of the cantilever beam (Fig. 3).

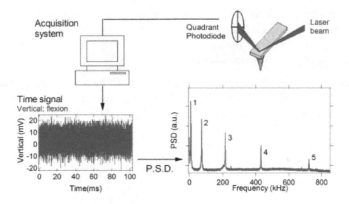

Fig. 3. Block diagram of the optical beam detection system. A typical power spectral density spectrum is shown. From (Malegori & Ferrini, 2010a).

Considering each flexural mode equivalent to a mass-spring system, the tip-sample interaction elastic constant is related to the frequency shift as $k_{ts} = -dF_{ts}/dz = 2k\Delta f/f_0$, see Eq. 12. The above relation holds if k_{ts} remains constant for the whole range of the displacements from the equilibrium position covered by the cantilever. This is usually true in the thermal regime since we are dealing with small oscillations (less than 0.2 nm) (Malegori & Ferrini, 2010a).

If this analysis is repeated at various separations from the surface, up to the jump-to-contact distance, the force gradient of the interaction dF_{ts}/dz is directly evaluated by the observed frequency shift of the PSD as a function of z.

From the same PSD, besides the force gradient, it is possible to measure the quality factor Q of the mode, that is determined by the relative width of the peaked structures corresponding to the oscillation eigenmodes of the cantilever ($\Delta\omega/\omega_0$). Q is usually dependent on the distance from the surface. Since the quality factor Q is connected to dissipation, important information on the tip-sample energy exchange can be retrieved.

With this techniques force gradients and quality factors on graphite and in air were measured by (Malegori & Ferrini, 2010a). It was found that the attractive force gradient data are well reproduced by a non-retarded van der Waals function in the form $HR/(3z^3)$ (H is the Hamaker constant and R the tip radius of curvature), up to the jump-to-contact distance which occurs at around 2 nm. In this distance range, Q is almost constant for the first and second flexural modes. This means that the interaction is conservative at distances greater than the jump-to-contact distance, the first flexural mode showing an evident decrease of the Q value just before the jump-to-contact. The dissipation mechanism related to this sharp transition is due to a local interaction of the tip apex with the surface.

In these experiments, the acquisition and storage of the photodiode time signal required tens of seconds at each tip-sample separation. This implies that the measurement at a single spatial location (one pixel of an image) may take minutes. The long measurements duration, besides the problems associated with the control of thermal drifts, is not practical for imaging purposes.

This difficulty stems from a precise characteristics of the Fourier analysis, which is devised for a stationary system i.e. the frequency spectrum is correctly correlated only with a temporally invariant physical system. For this reason, each measurement must be done in a quasi-steady state condition, requiring a long acquisition time.

As a consequence, the use of FT leaves aside many applications where the spectral content of the signal changes during the data collection. The spectral modifications are not revealed by FT, which only provides an average over the analyzed period of time and prevents correlating the frequency spectrum with the signal evolution in time. Clearly a different approach is needed to treat signals with a non-constant spectrum. In the next section, we describe a mathematical tool extremely useful to describe spectrally varying signals, the wavelet transforms.

3.2 Wavelet transforms

Perhaps one of the best ways to qualitatively appreciate the wavelet transform (WT) concept is an example. Consider a signal $f(t) = a \cos \varphi(t)$ with time varying phase $\varphi(t)$, where $\varphi(t) = \omega_0 t$ at negative times and $\varphi(t) = \omega_0 t + \alpha t^3$ at positive times (Fig. 4a). The instantaneous pulsation $\omega(t)$ is the derivative of the phase $\omega(t) = \varphi'(t)$. It is possible to see in Fig. 4b that WT analysis combines the time-domain and frequency-domain analysis so that the temporal evolution of each spectral component is determined. To confirm this, the calculated instantaneous pulsation (white line) is superposed to the signal amplitude obtained by the WT which is represented in color scale in the time-frequency plane. It is much like the concept of a musical score, where the pitch of a note (frequency) and its duration are displayed by the succession of the notes. In most cases, the wavelet analysis allows to extract accurately the instantaneous frequency information even for rapidly varying time series. To visualize the differences between the FT and WT, consider Fig. 4c. Since FT is a time invariant operator, only an average of the time-dependent spectrum is observed and the FT analysis is not able to correlate the frequency spectrum with the signal modifications in time. Instead, the wavelet transform represents the temporal trace in the time-frequency plane, providing the time dependence of both amplitude and frequency, see Fig. 4b.

To make a WT analysis, it is necessary to select an analyzing function $\Psi(t)$, called *mother wavelet*, (Mallat, 1999; Torrence & Compo, 1998). This wavelet must have zero mean and be localized in both time (unless a Fourier basis) and frequency space. An example is the Gabor wavelet, consisting of a plane wave modulated by a Gaussian

$$\Psi(t) = \frac{1}{(\sigma^2 \pi)^{1/4}} e^{\frac{t^2}{2\sigma^2} + i\eta t}$$

Here σ controls the amplitude of the Gaussian envelope, η the carrier frequency.

Dilations and translations of a mother wavelet $\Psi(t)$ generates the *daughter wavelets* as $\Psi_{s,d}(t) = \Psi(\frac{t-d}{s})$, where d is the delay and s the adimensional scale parameter. The wavelet dilations set by the scale parameter s are directly related to the frequency. The wavelet angular frequency at scale s is given by $\omega_s = \eta/s$. The associated frequency is $f_s = \omega_s/2\pi$. The wavelet translations set by the delay parameter d are obviously related to the time.

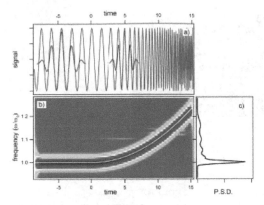

Fig. 4. Comparison between the Fourier transform and the wavelet transform analysis. a) The time signal, a cosine function for negative times and a cosine with quadratic chirp for positive times. Two daughter wavelet functions are superposed to the signal to show their localized similarity. b) Wavelet Transform of the temporal trace represented in a) showing the evolution of the frequency. The black line is the calculated instantaneous frequency. c) Fourier Transform (Power Spectral Density) of the signal represented in a). Only an average of the signal frequencies is observed. From (Malegori & Ferrini, 2010b).

The WT is defined as the convolution of the signal $f(t)$ with the daughter wavelets:

$$W(s,d) = \int_{+\infty}^{-\infty} f(t)\Psi_{s,d}^*(t)dt = \int_{+\infty}^{-\infty} f(t)\frac{1}{\sqrt{s}}\Psi^*(\frac{t-d}{s})dt$$

The square modulus of the wavelet coefficients $|W(s,d)|^2$, called the *scalogram*, represents the local energy density of the signal as a function of scale and delay (or equivalently frequency and time). The WT compares the signal with a daughter wavelet, measuring their similarity (see waveforms superposed to the signal in Fig. 4a). The coefficients $W(s,d)$ measure the similitude between the signal and the wavelet at various scales and delays. When the frequency of a daughter wavelet is close to that of the signal at a certain time, the WT amplitude reaches the maximum at that time and frequency position.

The instantaneous frequency of the signal is evaluated by the so called *wavelets ridges*, the maxima of the normalized scalogram (Mallat, 1999). When the signal contains several spectral lines whose frequencies are sufficiently apart, the wavelet ridges separate each of these components, as shown in Fig. 5.

3.2.1 The Heisenberg box

In time-frequency analysis both time resolution and frequency resolution have to be considered. As a consequence of the Heisenberg uncertainty principle, that holds for all wavelike phenomena, they cannot be improved simultaneously: when the time resolution increases, the frequency resolution degrades and vice versa. The time-frequency accuracy of the WT is limited by the time-frequency resolution of the corresponding mother wavelet. The WT resolution is confined in a box, the Heisenberg box, one dimension denoting the time

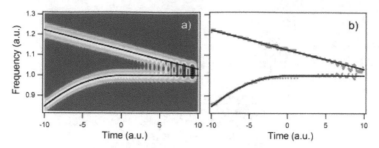

Fig. 5. a) wavelet transform of a signal that includes a linear chirp whose frequency decreases and a quadratic chirp whose frequency increases. b) The green points are the ridges calculated from the time-frequency topography. The black lines display the calculated instantaneous frequency of the linear and quadratic chirp. Note that the interference of the two spectral components destroys the ridge pattern.

resolution, the other the frequency resolution. The Heisenberg box delimits an area in the time-frequency plane over which different WT coefficients cannot be separated, providing a geometrical representation of the Heisenberg uncertainty principle. We adopt the commonly used definition of the measure of the uncertainty window Δ as the root-mean-square extension of the wavelet in the corresponding time or frequency space, (Malegori & Ferrini, 2010b),

$$\Delta_\xi^2 = \frac{\int_{-\infty}^{+\infty} \xi^2 \, |\Psi(\xi - \xi_0)|^2 \, d\xi}{\int_{-\infty}^{+\infty} |\Psi(\xi)|^2 \, d\xi}$$

where ξ_0 is a translation parameter and $\Psi(\xi)$ represents the Gabor mother wavelet, expressed either in time, $\xi = t$, or circular frequency, $\xi = \omega = 2\pi F$, $\Psi(\omega) = \text{FT}(\Psi(t))$.

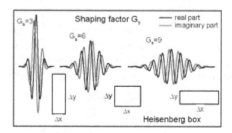

Fig. 6. Complex Gabor wavelet with different shaping factors. An increase of G_S corresponds to more oscillations. The "Heisenberg box" depicts the relationship between the time and frequency resolution, like the uncertainty principle in quantum mechanics (adapted from (Deng et al., 2005))

The time-frequency resolution of the analyzing Gabor mother wavelet, used in this work, is determined by the σ parameter. The Heisenberg box associated to the mother Gabor wavelet is given by a time resolution $\Delta_t = \sigma/\sqrt{2}$ and a frequency (or pulsation) resolution $\Delta_\omega = 1/(\sqrt{2}\sigma)$. When the wavelet is subject to a scale dilatation s, the corresponding resolution has size $\Delta_{s,t} = s\Delta_t$ along time and $\Delta_{s,\omega} = \Delta_\omega/s$ along frequency. The Heisenberg box centered at time t and angular frequency $\omega = 2\pi F$ is thus defined as

$$[t - \Delta_{s,t}, t + \Delta_{s,t}] \times [\omega - \Delta_{s,\omega}, \omega + \Delta_{s,\omega}]$$

The Heisenberg box area is four times $\Delta_{s,t}\Delta_{s,\omega} = 1/2$ (time resolution × frequency resolution) and is constant at all scales. The Gabor wavelet has the least spread in both frequency and time domain with respect to the choice of every other mother wavelet.

A single dimensionless parameter called the Gabor shaping factor $G_S = \sigma\eta$ controls the time-frequency localization properties of the Gabor mother wavelet (Deng et al., 2005). An increase of G_S means more oscillations under the wavelet envelope and a larger time spread, the frequency resolution being improved and the time resolution decreased, see Fig. 6

The resolution in time and frequency depends on s in such a way that the bandwidth-to-frequency ratio $\Delta_{s,f}/f_s$ is constant. In other words, in WT the window size changes adaptively to the frequency component by using shorter time supports to analyze higher frequency components and longer time supports to analyze lower frequency components. Therefore WT can accurately extract the instantaneous frequency from signals with wide spectral variation.

4. Wavelets meet Brownian motion: experimental results

Time-frequency analysis by wavelet transform is an effective tool to characterize the spectral content of signals rapidly varying in time. In this section we review the wavelet transform analysis applied to the thermally excited dynamic force spectroscopy to get insights into fundamental thermodynamical properties of the cantilever Brownian motion as well as giving a meaningful and intuitive representation of the cantilever dynamics in time and frequency caused by the tip-sample interaction forces.

Fig. 7 shows the time-frequency representation of the thermally excited *free* cantilever, i.e. a WT of the thermal noise of the cantilever flexural modes in air and at room temperature. The distinctive characteristic is the discontinuous appearance of the time-frequency traces, due to the discreteness and the statistical nature of the cantilever thermal excitation force F_{th}. To understand the appearance of the experimental trace and the dimensions of the observed bumps we need two concepts: the Heisenberg box and the *oscillator box*, that will be introduced in the next section.

The Heisenberg box is a visualization of the wavelet resolution (in Fig. 7 the vertical rectangles represents the Heisenberg boxes). Instead the oscillator box is related to the excitation and damping of the cantilever modes seen as damped harmonic oscillators, thus limiting the joint time-frequency response of the oscillator depending on resonant frequency and dissipation (in Fig. 7 the horizontal rectangles represents the oscillator boxes).

It is remarkable that the dimension of the bumps observed in the experimental time-frequency traces are accounted for quantitatively by the dimensions of the boxes mentioned above not only for different flexural modes, but also for contact modes and torsional modes.

Moreover, we would like to emphasize that one of the advantages of the wavelet analysis lies in the possibility to carry out measurements across the jump-to-contact transition without

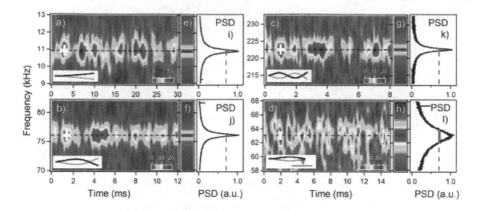

Fig. 7. a)-c) Wavelet transform of the free cantilever thermal vibrations for the three lower flexural eigenmodes. The wavelet coefficients $|W(f,t)|$ are coded in color scale. The horizontal white rectangles represent the damped oscillator boxes, the vertical rectangles the Heisenberg boxes. The dashed lines correspond to the resonant frequencies measured from the power spectral density. d) Same as a)-c) for the clamped cantilever exerting a positive load of approximately 1.1 nN on the surface. The rectangle on the left is the Heisenberg box, that on the right the damped oscillator box. i)-k) Square root of the normalized power spectral density of the free cantilever Brownian motion zoomed on the lower three resonant frequencies. The blue line is the frequency linewidth $\Delta f = f_0/Q$. The corresponding color scale plots are displayed in e)-g). l) Same as i)-k) for the clamped cantilever. The corresponding colorscale plot is displayed in h). From (Malegori & Ferrini, 2011a).

interruption (Malegori & Ferrini, 2011a), providing information on the long- and short-range adhesion surface forces. Tip-sample van der Waals interaction, adhesion forces, friction and elastic properties of the surface can be measured in few 10's of milliseconds, a time compatible with practical Dynamic Force Spectroscopy imaging.

4.1 The oscillator box

The dynamics of a free cantilever in air can be reasonably modeled as an harmonic oscillator with viscous dissipation. If no driving forces are applied, the cantilever is excited by stochastic forces whose amplitude are connected to the ambient temperature (the thermal reservoir) by the fluctuation-dissipation theorem, see Sec. 2.2. Microscopically, the external thermal force F_{th} can be thought as the action of random thermal kicks (uncorrelated impulsive forces), with a white frequency spectrum. This thermal force induces cantilever displacements from the equilibrium position, that show a marked amplitude enhancement in correspondence of the flexural eigenfrequencies.

The motion of a damped harmonic oscillator after an impulsive force excitation constitutes a simple model to describe the cantilever dynamics after a single thermal kick. When the cantilever has a thermally activated fluctuation, each flexural mode responds as a damped harmonic oscillator whose equation of motion is $\ddot{x} + \omega_0/Q\dot{x} + \omega_0^2 x = 0$ where x is the oscillation amplitude (the dots denotes the derivative with respect to time), Q the quality factor and ω_0 the resonance frequency (Albrecht et al., 1991; Demtröder, 2003). Considering for simplicity the initial conditions $x(0) = x_0$, $\dot{x}(0) = 0$ and assuming $Q \gg 1$, the solution is an exponentially decaying amplitude oscillating at the resonance frequency: $x = x_0 e^{-\omega_0 t/(2Q)} cos(\omega_0 t)$.

The energy $E(t)$ associated to the oscillator is proportional to the maximum of $\dot{x}^2$ and from the above relations we see that, in case of small damping, the associated exponential energy decay time is $\tau = Q/\omega_0$. The spectral energy density of the damped oscillator ($L(\omega)$) is proportional to the square modulus of the Fourier transform of $x(t)$, $L(\omega) = |FT(x(t))|^2$. Under the assumption $Q \gg 1$, $L(\omega)$ is well approximated by a Lorentzian with a full width at maximum height of $\Delta\omega = 2\pi\Delta f = 1/\tau$.

Since the cantilever is first thermally excited and then damped to steady state by random forces that act on a much smaller time scale than its oscillation period, the characteristic response time for an isolated excitation/decay event cannot be smaller than 2τ, with an associated Lorentzian full width at half maximum of $\Delta\omega$.

From the above reasoning, it is natural to introduce the *damped oscillator box*, a geometrical representation of the extension in the time-frequency plane of the wavelet coefficients associated to a single excitation/decay event, centered at time t and frequency ω, defined as

$$[t - \tau, t + \tau] \times [\omega - \Delta\omega/2, \omega + \Delta\omega/2]$$

The damped oscillator box, contrary to the Heisenberg box, does not represent a limitation in resolution due to the wavelet choice, but a physical representation of the damped oscillator time-frequency characteristics. It is important to note that the ultimate resolution limitations imposed by the Heisenberg box associated with the analyzing wavelet could prevent the observation of the true dimensions of the damped oscillator box.

Although the free oscillation modes have very different resonant frequencies, mode shapes and relaxation times, the discrete structures near resonance have dimensions close to the respective damped oscillator boxes, within the wavelet resolution, suggesting that these structures are correlated with *single* thermal excitation events.

From the PSD shown in (Fig. 7 i,j,k), f_0 and Q are obtained from a Lorentzian fit of the resonance peaks (see Tab. 1) and the damped oscillator box dimensions are calculated. For the first, second and third free flexural eigenmodes $\tau_\gamma = 0.93, 0.38, 0.23$ ms and $\Delta\omega = 1.1, 2.7, 4.4$ kHz respectively. The Heisenberg box dimensions for the same modes are $\Delta_t = 0.51, 0.18, 0.11$ ms and $\Delta_\omega = 3.9, 11, 18$ kHz respectively. In this case, $\Delta_t < 2\tau_\gamma$, and $\Delta_\omega > \Delta\omega$. As a consequence, the temporal width of the smaller structures is about $2\tau_\gamma$, while their spectral width is determined by the wavelet frequency resolution. The temporal width of the structures is independent on the time resolution of the wavelet (provided that it is smaller than $2\tau_\gamma$), indicating that we are observing a real physical feature not related to the choice of the wavelet representation.

A different case is represented by the thermal oscillations of the surface-coupled cantilever, shown in Fig. 7d, where the tip is clamped and exerts a constant force of 1.1 nN on the surface during the measurement. In this case the tip does not oscillate and the temporal trace recorded by the optical beam deflection system is proportional to the slope at the cantilever end. Describing this motion as a damped harmonic oscillator, we have $\tau_\gamma = 0.082$ ms and $\Delta\omega = 12$ kHz. The Heisenberg box values are $\Delta_t = 0.27$, ms and $\Delta_\omega = 7.5$ kHz. Due to the lower Q factor the spectral width is wider and the decay time smaller so that, contrary to the free cantilever case, $\Delta_t > 2\tau_\gamma$ and $\Delta_\omega < \Delta\omega$. Now the temporal width is limited by the wavelet temporal resolution while the frequency width is that of the damped oscillator. Therefore, WT describes more easily the time decay of a single thermal excitation event in high-Q environments and its frequency linewidth in low-Q environments.

We conclude observing that the discrete time-frequency small structures seen in the time-frequency representation, related to the cantilever excitation and decay to steady state by a single thermal fluctuation event, can be regarded as a visualization of the consequences of the fluctuation-dissipation theorem.

mode	f_1	f_2	f_3	f_4	f_{c1}	f_{c2}
f_n (kHz)	10.908	76.09	222.6	444.4	62.80	195.7
Q_n	63	180	320	470	32	89
f_n/f_1 exp.	1	6.97	20.4	40.7	5.75	17.9
f_n/f_1 teo.	1	6.26	17.5	34.4	4.38	14.2

Table 1. Free (f_1–f_4) and clamped (f_{c1}–f_{c2}) cantilever resonant frequencies and quality factors. The measured ratio between the frequencies of the higher modes with the first one is compared with the theoretical prediction of (Butt & Jaschke, 1995).

4.2 Force spectroscopy

The wavelet analysis is applied to the force-distance curves taken with the cantilever subject to thermal fluctuations while approaching the surface. Fig. 8 shows the scalogram of a 40 ms sampling of the cantilever Brownian motion around its instantaneous equilibrium position while the piezo scanner is displaced at constant velocity to move the tip towards the surface, until it jumps to contact.

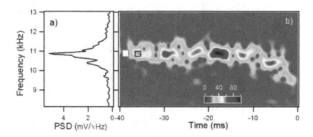

Fig. 8. a) Power Spectral Density of the Brownian motion of the first flexural mode as the tip approaches the surface at constant velocity (9nm/40ms=225 nm/s). b) Wavelet transform of the same temporal trace. The wavelet coefficients $|W(f,t)|$ are coded in color scale. The origin of the time axis corresponds to the instant when the jump to contact occurs. The white box on the left side is the Heisenberg box, the open box delimited by black lines is the damped oscillator box. From (Malegori & Ferrini, 2010b)

Since the cantilever is at room temperature T, its mean potential energy $1/2k\langle x^2\rangle$ is equal to $1/2k_BT$ by the equipartition theorem. This thermal force induces cantilever displacements from the equilibrium position, that show a marked amplitude enhancement in correspondence of the first flexural eigenfrequency. It is clear that the thermally driven eigenfrequency is affected by the tip-sample interaction forces in a small time interval prior to the jump-to-contact transition, causing a frequency decrease, as shown by the wavelet analysis in Fig. 8b. The PSD of the same temporal trace used for the WT, reported in Fig. 8a, shows a linewidth comparable to the frequency indetermination of the Heisenberg box of the WT and a structure at low frequency that is reminiscent of the interaction with the surface, when for a short time the cantilever frequency is lowered.

4.2.1 Force spectroscopy analysis

As observed previously, the instantaneous frequency is evaluated by the wavelet ridges, the local maxima points of the normalized scalogram. In order to reduce noise effects, only maxima above a threshold are considered (see the schematic representation in the inset of Fig. 9).

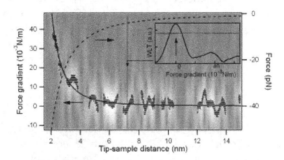

Fig. 9. Force gradient versus tip-sample distance for the first flexural mode near the jump-to-contact. The wavelet ridges provide the instantaneous frequencies within the limits of the scalogram resolution. The wavelet ridges are the local maxima of the normalized scalogram above a specified threshold, as schematically shown in the inset. The threshold is represented by a horizontal line and the maximum point is indicated by an arrow for a vertical cut of the data at constant tip-sample distance. The WT of Fig. 8 is represented in gray scale on the background together with its ridges (black points). The continuous black line is an Hamaker-like force gradient function fitted to the wavelet ridges, the dashed line the force calculated by mathematical integration. From (Malegori & Ferrini, 2010b)

The first flexural mode frequency shift near the surface (Fig. 8b) provides a complete force-distance curve using the wavelet ridges. From the instantaneous frequency shift the gradient of the tip-sample interaction forces (dF_{ts}/dz) is retrieved, using the relations previously reported. The time scale is converted into the tip-sample separation by taking into account the piezoscanner velocity and the cantilever static deflection, to obtain a complete force gradient versus distance curve (Fig. 9).

The gradient data from WT ridges are well fitted by a non-retarded van der Waals function in the form $HR/3z^{-3}$, with $HR = 1.2 \times 10^{-27}$ Jm. Using the typical values of H in graphite ($H = 0.1$ aJ), the tip radius is evaluated as $R = 12$ nm, in good agreement with the nominal radius of curvature given by the manufacturer ($R = 10$ nm). To promote this technique from proof of principle to a measurement of the Hamaker constant with a good lateral resolution, a thorough characterization of the tip radius of curvature is needed.

The whole force curve is acquired in less than 40 ms, a time significantly shorter than that usually needed for force versus distance measurements. With an optimization of the electronics and reduction of dead times in the acquisition process, it would be possible to acquire images with the complete information on force gradients and topography compatible with 1-30 ms/pixel data acquisition times required for practical DFS imaging.

4.2.2 Contact dynamic force spectroscopy

The jump-to-contact (JTC) transition is accompanied by a high-amplitude damped oscillation of the clamped cantilever started by the impact of the tip on the surface, visible immediately after the transition. In this case the tip, attracted by the short-range adhesion forces, behaves like a nano-hammer. The wavelet transform can be carried out across the JTC transition without interruption and the oscillations induced by the JTC event are shown in the wavelet representation as a big bump in the time-frequency space (Fig. 10a). From the temporal traces we estimate that the cantilever takes approximately 10 μs to collapse into the new state with a clamped end (Fig. 10b), a duration shorter than the system oscillation period that can be considered as instantaneous on the cantilever typical timescales. The changed boundary condition (from free to clamped cantilever end) produces a sudden variation of the flexural resonant frequencies.

In the experimental data, the time scale is converted to cantilever deflection scale taking into account the piezotube movement and the position of the surface deduced by the deflection vs distance curve (the solid-liquid interface). Negative deflection means that the beam is bent toward the sample. The load of the tip on the sample is directly calculated as $F_{load} = kx$ where k and x are the cantilever elastic constant and static deflection, respectively. In this case the loading is negative since the contact is kept by adhesion force that opposes the elastic force of the bent cantilever. The transient frequency analysis allows retrieval of the oscillator Q factor by measuring the ratio of the oscillation frequency to the frequency width of the initial high-amplitude damped oscillation. Since the Heisenberg box dimension of the analyzing wavelet is 0.27 ms $\times$ 1.2 kHz and the frequency width is of the order of $\Delta f = 2$kHz, the frequency width is not limited by the wavelet resolution. With a central frequency of about 60 kHz, the Q factor is obtained as $Q = f/\Delta f = 60$kHz$/(2$kHz$)=30$. This estimate is quite consistent with the Q factor found in contact mode under static loading (Table 1), confirming that the physical oscillator (the cantilever) has the same dissipation dynamics in the various interaction-force regimes (negative and positive loading) after JTC.

The resonant frequency has an evident increase caused by the decrease in the adhesion forces due to cantilever moving towards the surface at constant velocity, a behavior reproducible in all our measurements. The frequency shift is related to the total force (adhesion plus elastic force) that decreases as the cantilever negative deflection decreases towards zero. This transient behavior could not be captured with standard or non-dynamical techniques.

It is possible to continue the contact mode WT analysis increasing the load up to higher positive values. Fig. 11 shows the ridge analysis of the entire spectroscopy curve. After the transient at negative loading described above, the cantilever passes through the zero-load

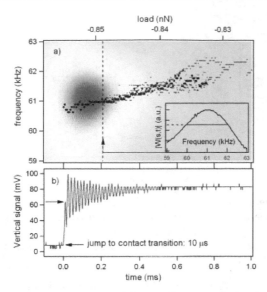

Fig. 10. a) Time-frequency representation of the cantilever evolution immediately after the jump to contact. Resonant frequency is about six times the free cantilever frequency due to changed boundary condition (clamped end). The big bump is due to the cantilever oscillations upon tip impact with the surface. The instantaneous frequency versus load is provided by the wavelet ridges analysis. Three different measurements (black, gray and light grey points) are shown to demonstrate reproducibility. In the inset: normalized wavelet coefficient $|W(f,t)|$ along the vertical dashed line. b) The temporal evolution of the clamped cantilever oscillations immediately after the jump to contact transition. It is evidenced the short time interval (approximately $10\mu s$) for the cantilever to collapse into contact. From (Malegori & Ferrini, 2011a)

neutral point, where it is not deflected, and then continues with increasing positive load on the surface. The frequency shift can be followed starting from the very beginning of the cantilever interaction with the surface and with good temporal resolution. The single measure is taken in approximately 100 ms. With an appropriate analysis, it would be possible to study in detail both the adhesion forces dynamics of the cantilever (Espinosa-Beltrán et al., 2009; Yamanaka & Nakano, 1998) and the elasticity parameters (e.g the Young's modulus) from the contact region (Hertz contact dynamics) (Dupas et al., 2001; Espinosa-Beltrán et al., 2009; Rabe et al., 1996; Vairac et al., 2003).

As a final comment to this section, we stress that the wavelet transform approach could provide quantitative information on the surface elastic properties especially when low force regimes are needed, i.e. on softer samples such as biological or polymer surfaces.

4.3 Torsional modes

The torsional modes the cantilever oscillate about its long axis and the tip moves nearly parallel to the surface, being sensitive to in-plane forces. As a consequence, the eigenfrequency of the torsional modes depends only on the lateral stiffness of the sample, serving as shear

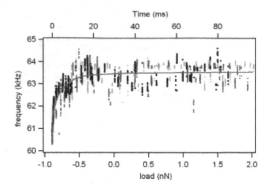

Fig. 11. Resonant frequency versus load of the first flexural contact mode, spanning the negative and positive loading regime. The instantaneous frequency versus load is provided by the wavelet ridges analysis. Three different measurements (black, gray and light gray points) are shown to demonstrate reproducibility. The continuous line is a guide to the eye. From (Malegori & Ferrini, 2011a)

stiffness sensors. An increasing shear stiffness increases the lateral spring constant and consequently the resonant frequency of the system (Drobek et al., 1999). We study the spectra of thermally excited torsional modes of the cantilever as the tip approaches a graphite surface in air (Malegori & Ferrini, 2011b). Since we are interested in exploring what happens immediately after the JTC transition, the forces that predominate in this regime are the attractive adhesion forces.

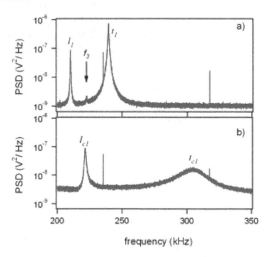

Fig. 12. a) Thermal power spectral density of the cantilever torsional fluctuations zoomed on the first torsional (t_1) and first lateral (l_1) resonance peaks. The arrow points at a small contribution from the third flexural mode (f_3) at 223 kHz. b) Same as a) but with the tip in contact with the sample at constant negative load (-0.5 nN). From (Malegori & Ferrini, 2011b).

The power density spectrum of the free cantilever first torsional modes is reported in Fig. 12a. The peak at 239.4 kHz with $Q=310$ is the first torsional mode (t_1). The mode at 210.2 kHz with $Q=590$ is the first lateral bending mode (l_1) (Espinosa-Beltrán et al., 2009). The lateral modes are cantilever in-plane flexural modes, in contrast with the usual out-of-plane flexural modes. In the spectrum is also visible a minute contribution from the third flexural mode at 222 kHz. When the tip is brought close to the sample, the capillary forces attract the tip to the HOPG surface until the JTC transition occurs (Luna et al., 1999). Due to the modified mechanical boundary conditions, the cantilever end is no longer free. A clear shift of the torsional and lateral contact mode resonances is detected under a negative static load of -0.5 nN, Fig. 12b. The first contact torsional mode resonance frequency increases to 305.2 kHz with $Q=14$; the contact lateral mode resonance frequency increases to 221.7 kHz with $Q=200$. In both cases the dissipation increases for contact modes, particularly for the first torsional eigenmode.

The torsional resonance variation of the thermally excited cantilever can be followed across the JTC transition with the wavelet transforms, as shown in Fig. 13. The JTC transition is located at time zero, separating the negative times of the free cantilever evolution, from the positive times of the clamped cantilever evolution. Note that the long-range forces and capillary phenomena that usually interfere with the oscillations of the flexural modes (Jesse et al., 2007; Malegori & Ferrini, 2010a;b; Roters & Johannsmann, 1996) do not perturb the much stiffer torsional free modes until jump-to-contact. The lateral mode frequency displays a very sharp frequency shift at JTC and remains fairly constant immediately after. Instead, the torsional contact mode shows a detectable and continuous frequency increase after JTC, caused by the tip interaction with the graphite surface.

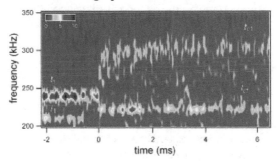

Fig. 13. Wavelet transform of the cantilever thermal torsional oscillation across the jump-to-contact transition, showing the evolution of the first free torsional mode t_1 into the contact torsional mode t_{c1} and the first free lateral mode l_1 into the contact lateral mode l_{c1}. The wavelet coefficients $|W(f,d)|$ are coded in color scale. The origin of the time axis is at the jump-to-contact onset. Both modes have an evident shift as the tip is attracted on the surface. From (Malegori & Ferrini, 2011b).

It is worth noting the different appearance of the torsional mode frequency structure before and after JTC in Fig. 13. It is evident a sudden increase of the frequency width (and a corresponding decrease of the time width) of the time-frequency trace passing through the JTC point (time zero), that can be qualitatively explained as a sudden increase in dissipation caused by the interaction with the surface. This demonstrates that there is not a smooth transition during the JTC between the free and contact oscillations.

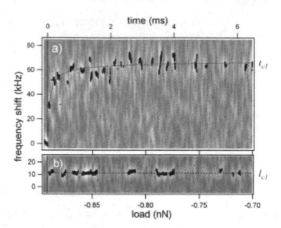

Fig. 14. a) Frequency shift with respect to the free resonant frequency of the first contact torsional mode t_{c1} versus the tip load. The ridges (black points) of the wavelet in Fig. 13, represented in gray scale on the background, provide the instantaneous frequency within the limit of the scalogram resolution. The continuous vertical line at time zero corresponds to the jump-to-contact onset. The dotted line is a guide to the eye. b) Ridges of the frequency shift with respect to the free resonant frequency of the first contact lateral mode l_{c1} versus the tip load. The dotted line is the lateral contact resonant frequency. From (Malegori & Ferrini, 2011b).

Taking into account the piezo-scanner vertical velocity, it is possible to obtain a linear relation between time and cantilever deflection, allowing calculation of the contact loading force of the tip on the surface. The frequency evolution is provided by the wavelet ridges, showing the instantaneous frequencies within the transform resolution limits in Fig. 14 as black points. Using the wavelet ridges, after JTC the time-frequency representation is transformed into a contact-interaction-force vs frequency-shift representation.

Immediately after JTC the force acting on the cantilever is negative (negative loading). In this case the tip is acted on by adhesion forces that attracts the tip towards the surface. The frequency shift of resonance frequencies with respect to the free cantilever oscillations is thus caused by the decrease in strength of adhesion forces, a transient that could not be easily captured with standard or non-dynamical techniques.

With a suitable model this technique could allow a thorough measurement of the adhesion force properties (Drobek et al., 2001; Espinosa-Beltrán et al., 2009; Yamanaka & Nakano, 1998). Analytical and numerical models describing the free cantilever-vibration as well as the contact-resonances are well known and provide quantitative evaluation when complete contact-resonance spectra are measured. The contact-resonance frequencies of the cantilever are linked to the tip-sample contact stiffness, which depends on the elastic indentation moduli of the tip and the sample and on the shape of the contact. The spatial resolution depends on the tip-sample contact radius, which is usually in the range from 10 to 100 nm. Lateral stiffness determined from the contact resonant frequencies of the first torsional vibration obtained from noise spectra have already been investigated in (Drobek et al., 2001) using quasi-static force curve cycles. The improvement provided by WT analysis is related to the time required to detect the frequency shift vs load curve which is in the order of few ms. This acquisition time is significantly shorter than that of the quasi-static techniques and compatible with the development of real-time measurement.

To analyze a further example of the interplay between the wavelet resolution and the thermal excitation, we now consider two extreme cases: the free cantilever and the clamped cantilever with positive loading.

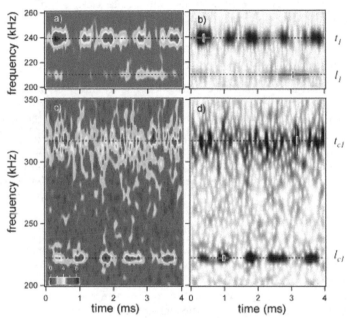

Fig. 15. a) Wavelet transform of the free cantilever thermal fluctuations of the first lateral (l_1) and first torsional (t_1) mode. The wavelet coefficients $|W(f,d)|$ are coded in color scale. The dotted lines are centered on the resonant frequencies of the modes. b) Same image as in a) but coded in saturated gray scale to appreciate the shape of the discontinuous structures. c) and d) Same as a) and b) but for the first lateral (l_{c1}) and first torsional (t_{c1}) contact mode at constant positive load of approximately 1.6 nN. In b) and d) the red rectangles with a white border represent the Heisenberg boxes. The red rectangles represent the damped oscillator boxes. From (Malegori & Ferrini, 2011b).

The wavelet transform of the free thermal oscillations of the cantilever detected by the left-right sections of the quadrant photodiode shows the time evolution of the first torsional mode and the first lateral mode (see Fig. 15a). When the Q factor of a mode is high (see Tab. 2), the corresponding frequency linewidth is small. In this case the frequency resolution of a wavelet may be not sufficient to resolve the intrinsic linewidth of the mechanical resonance. The Heisenberg box dimensions are 0.050 ms × 6.35 kHz for the first torsional mode (t_1) and 0.057 ms × 5.6 kHz for the first lateral mode (l_1). The damped oscillator boxes for the same modes are 0.41 ms × 0.77 kHz (t_1) and 0.90 ms × 0.35 kHz (l_1). Thus the frequency width of the time-frequency distribution is limited by the wavelet resolution (the frequency width of the Heisenberg box) which is much higher than the frequency width of the oscillator box (see

mode	t_1	t_2	t_{c1}	t_{c2}	l_1	l_{c1}
Frequency (kHz)	239.4	763.4	305	782	210.1	221.6
Quality factor (Q)	310	570	14	120	590	200
Frequency ratio exp.	22.0	70.0	28.0	71.7	19.3	20.3
Frequency ratio teo.	23.4	70.1	–	–	19.0	–

Table 2. Comparison between measured and calculated (Butt & Jaschke, 1995; Espinosa-Beltrán et al., 2009) free cantilever resonant frequencies. The theoretical results are expressed as ratios with respect to the first flexural frequency, $f_1 = 10.908$ kHz. The Q factors are measured from the power density spectra. t and l refers to the free torsional and lateral eigenmodes. t_c and l_c refers to the contact torsional and lateral eigenmodes. In this case the contact measurements refers to a negative load on the tip of -0.5 nN.

Fig. 15b). On the other hand, a high Q implies a long decay time associated to the oscillator energy. In this case the time associated to the damped oscillator box is larger than the temporal wavelet resolution, i.e. the time width of the oscillator box is larger than the time width of the Heisenberg box. In Fig. 15b the oscillator boxes (red) and the Heisenberg boxes (red with a white border), have been superposed on the time-frequency representation of the wavelet coefficients. In this case the Heisenberg box, i.e the wavelet resolution, limits the frequency width while the temporal extension of the structures is similar to the oscillator box time width. Such structures can be interpreted as the cantilever excitation and decay to steady state after a single thermal fluctuation event (Malegori & Ferrini, 2011a).

Fig. 15c shows the contact cantilever vibrations after JTC at a static positive load of the tip on the graphite surface of approximately 1.6 nN. The Q factor of the first torsional contact mode (t_{1c}) decreases and the oscillator box re-shapes accordingly, reducing the damping time and increasing its frequency width (Fig. 15d). We found the Q factors of the contact modes to be almost independent from the tip loading in the studied range and similar to those reported in Tab. 2 for negative loading. In this case the Heisenberg box dimensions are 0.053 ms $\times$ 6.0 kHz for the first torsional contact mode (t_{c1}) at 316.72 kHz and 0.075 ms $\times$ 4.2 kHz for the first lateral mode (l_{c1}) at 221.6 kHz. The damped oscillator boxes for the same modes are 0.013 ms $\times$ 23.3 kHz (t_{c1}) and 0.28 ms $\times$ 1.25 kHz (l_{c1}). As can be seen from the data reported above, the frequency resolution of the wavelet for the mode t_{c1} is sufficient to reconstruct the linewidth profile of the time-frequency trace, i.e. the Heisenberg box spectral width is smaller than the frequency width of the oscillator box. In contrast to the other modes, the time resolution of the wavelet does not allow to follow the temporal evolution of the single thermal excitation, because the time width of the oscillator box is smaller than the corresponding Heisenberg box dimension.

5. Conclusions

The wavelet analysis applied to dynamic AFM is especially useful in capturing the temporal evolution of the spectral response of the interacting cantilever. In this respect, the applications of wavelet analysis to the thermally driven cantilever to detect forces rapidly and continuously varying across the jump-to-contact transition must be seen as just examples and do not exhaust all the possible utilizations. Traditional AFM techniques enable the construction of the spectral response by modifying the cantilever interaction step by step. However, in this way, it is not possible to analyze transients. Instead, the wavelet analysis allows detection of transient spectral features that are not accessible through steady state techniques. Moreover, the ability to capture the relevant spectral evolution in a time frame of tens of milliseconds enables surface chemical kinetics or surface force modification to be tracked in real time with dynamic force spectroscopy. More fundamentally, the wavelet transforms highlight the thermodynamic characteristics of the cantilever Brownian motion, enabling the tip-sample fluctuation-dissipation interactions to be investigated. In conclusion, although the results reviewed in the present work must be considered as preliminary, the proposed technique is interesting in view of its simplicity and connection with fundamental thermodynamic quantities.

6. Acknowledgments

This work has been supported by the Università Cattolica del Sacro Cuore through D.2.2 and D.3.1 grants.

7. References

Albers, B. J., Schwendemann, T. C., Baykara, M. Z., Pilet, N., Liebmann, M., Altman, E. I. & Schwarz, U. D. (2009). Three-dimensional imaging of short-range chemical forces with picometre resolution, *Nature Nanotech.* 4: 307.

Albrecht, T. R., Grütter, P., Horne, D. & Rugar, D. (1991). Frequency modulation detection using high-Q cantilevers for enhanced force microscope sensitivity, *J. Appl. Phys.* p. 668.

Andreeva, N., Bassi, D., Cappa, F., Cocconcelli, P., Parmigiani, F. & Ferrini, G. (2010). Nanomechanical analysis of clostridium tyrobutyricum spores, *Micron* 41: 945.

Binnig, G., Quate, C. C. & Gerber, C. (1986). Atomic force microscope, *Phys. Rev. Lett.* 56: 930.

Binnig, G., Rohrer, H., Gerber, C. & Weibel, E. (1982). Surface studies by scanning tunneling microscopy, *Phys. Rev. Lett.* 49: 57.

Braga, P. C. & Ricci, D. (2004). *Atomic force microscopy: biomedical methods and applications*, Humana Press Inc.

Butt, H. J., Cappella, B. & Kappl, M. (2005). Force measurements with the atomic force microscope: Technique, interpretation and applications, *Surf. Sci. Reports* 59: 1.

Butt, H.-J. & Jaschke, M. (1995). Calculation of thermal noise in atomic force microscopy, *Nanotechnology* 6: 1.

Callen, H. B. & Greene, R. F. (1952). On a theorem of irreversible thermodynamics, *Phys. Rev.* 86: 702.

Cappella, B. & Dietler, G. (1999). Force-distence curves by atomin force microscopy, *Surf. Sci. Reports* 34: 1.

Cleveland, J. P., Schäffer, T. E. & Hansma, P. K. (1995). Probing oscillatory hydration potentials using thermal-mechanical noise in an atomic-force microscope, *Phys. Rev. B* 52: R8692.

Crittenden, S., Raman, A. & Reifenberger, R. (2005). Probing attractive forces at the nanoscale using higher-harmonic dynamic force microscopy, *Phys. Rev. B* 72: 235422.

Demtröder, W. (2003). *Laser Spectroscopy*, third edn, Springer.

Deng, Y., Wang, C., Chai, L. & Zhang, Z. (2005). Determination of Gabor wavelet shaping factor for accurate phase retrieval with wavelet-transform, *Appl. Phys. B* 81: 1107.

Drobek, T., Stark, R. W., Gräber, M. & Heckl, W. M. (1999). Overtone atomic force microscopy studies of decagonal quasicrystal surfaces, *New J. Phys.* 1: 15.1.

Drobek, T., Stark, R. W. & Heckl, W. M. (2001). Determination of shear stiffness based on thermal noise analysis in atomic force microscopy: Passive overtone microscopy, *Phys. Rev. B* 64: 045401.

Dupas, E., Gremaud, G., Kulik, A. & Loubet, J.-L. (2001). High-frequency mechanical spectroscopy with an atomic force microscope, *Rev. Sci. Instrum.* 72: 3891.

Espinosa-Beltrán, F. J., Geng, K., Muñoz Saldaña, J., Rabe, U., Hirsekorn, S. & Arnold, W. (2009). Simulation of vibrational resonances of stiff AFM cantilevers by finite element methods, *New J. Phys.* 11: 083034.

Fukuma, T. (2010). Water distribution at solid/liquid interfaces visualized by frequency modulation atomic force microscopy, *Sci. Technol. Adv. Mater.* 11: 033003.

Fukuma, T., Ueda, Y., Yoshioka, S. & Asakawa, H. (2010). Atomic-scale distribution of water molecules at the mica-water interface visualized by three-dimensional scanning force microscopy, *Phys. Rev. Lett.* 104: 016101.

Garcia, R. (2010). *Amplitude modulation atomic force spectroscopy*, first edn, Wiley-VCH.

Garcia, R. & Perez, R. (2002). Dynamic afm methods, *Surf. Sci. Reports* 47: 197.

Giessibl, F. J. (2003). Advances in atomic force microscopy, *Rev. Mod. Phys.* 75: 949.

Gillespie, D. T. (1993). Fluctuation and dissipation in brownian motion, *Am. J. Phys.* 61: 1077.

Gillespie, D. T. (1996). The mathematics of Brownian motion and Johnson noise, *Am. J. Phys.* 64: 225.

Gross, L., Mohn, F., Moll, N., Liljeroth, P. & Meyer, G. (2009). The chemical structure of a molecule resolved by atomic force microscopy, *Science* 325: 1110.

Heinz, W. F., Antonik, M. D. & Hoh, J. H. (2000). Reconstructing local interaction potentials from perturbations to the thermally driven motion of an atomic force microscope cantilever, *J. Phys. Chem. B* 104: 622.

Hölscher, H., Allers, W., Schwarz, U. D., Schwarz, A. & Wiesendanger, R. (1999). Determination of tip-sample interaction potentials by dynamic force spectroscopy, *Phys. Rev. Lett.* 83: 4780.

Hutter, J. L. & Bechhoefer, J. (1993). Calibration of atomic-force microscope tips, *Rev. Sci. Instrum.* 64: 1868.

Jandt, K. (2001). Atomic force microscopy of biomaterials surfaces and interfaces, *Surf. Sci.* 491: 303.

Jena, B. P. & Hörber, J. K. H. (eds) (2002). *Atomic Force Microscopy in Cell Biology*, Academic Press.

Jesse, S., Kalinin, S. V., Proksch, R., Baddorf, A. P. & Rodriguez, B. J. (2007). The band excitation method in scanning probe microscopy for rapid mapping of energy dissipation on the nanoscale, *Nanotechnology* 18: 435503.

Kopniczky, J. (2003). *Nanostructures Studied by Atomic Force Microscopy*, Acta Universitatis Upsaliensis, Uppsala Sweden.

Koralek, D. O., Heintz, W. F., Antonik, M. D., Baik, A. & Hoh, J. H. (2000). Probing deep interaction potentials with white-noise-driven atomic force microscope cantilevers, *Appl. Phys. Lett.* 76: 2952.

Lantz, M. A., Hug, H. J., Hoffmann, R., van Schendel, P. J. A., Kappenberger, P., Martin, S., Baratoff, A. & Güntherodt, H. J. (2001). Quantitative measurement of short-range chemical bonding forces, *Science* 291: 2580.

Luna, M., Colchero, J. & Baró, A. M. (1999). Study of water droplets and films on graphite by noncontact scanning force microscopy, *J. Phys. Chem. B* 103: 9576.

Malegori, G. & Ferrini, G. (2010a). Tip-sample interactions on graphite studied in the thermal oscillation regime, *J. Vac. Sci. Technol. B* 28: C4B18.

Malegori, G. & Ferrini, G. (2010b). Tip-sample interactions on graphite studied using the wavelet transform, *Beilstein J. Nanotechnol.* 1: 172.

Malegori, G. & Ferrini, G. (2011a). Wavelet transforms to probe long- and short-range forces by thermally excited dynamic force spectroscopy, *Nanotechnology* 22: 195702.

Malegori, G. & Ferrini, G. (2011b). Wavelet transforms to probe the torsional modes of a thermally excited cantilever across the jump-to-contact transition: Preliminary results, *e-J. Surf. Sci. Nanotech.* 9: 228.

Mallat, S. G. (1999). *A Wavelet Tour of Signal Processing*, Academic Press.

Martin, Y., Williams, C. C. & Wickramasinghe, H. K. (1987). Atomic force microscope-force mapping and profiling on a sub 100- scale, *J. Appl. Phys.* 61: 4723.

Mironov, V. L. (2004). *Fundamentals of scanning probe microscopy*, The Russian Academy of Sciences, Nizhniy Novgorod Russia.

Morita, S., Giessibl, F. J. & Wiesendanger, R. (eds) (2009). *Noncontact Atomic Force Microscopy*, Vol. II, Springer, Berlin.

Morita, S., Wiesendanger, R. & Meyer, E. (eds) (2002). *Noncontact Atomic Force Microscopy*, Springer, Berlin.

Palacios-Lidón, E. & Colchero, J. (2006). Quantitative analysis of tip-sample interaction in non-contact scanning force spectroscopy, *Nanotechnology* 17: 5491.

Paolino, P. & Bellon, L. (2009). Frequency dependence of viscous and viscoelastic dissipation in coated micro-cantilevers from noise measurement, *Nanotechnology* 20: 405705.

Rabe, U., Janser, K. & Arnold, W. (1996). Vibrations of free and surface-coupled atomic force microscope cantilevers: Theory and experiment, *Rev. Sci. Instrum.* 67: 3281.

Radmacher, M., Fritz, M., Kacher, C. M., Cleveland, J. P. & Hansma, P. K. (1996). Measuring the viscoelastic properties of human platelets with the atomic force microscope, *Biophys. J.* 70: 556.

Roters, A. & Johannsmann, D. (1996). Distance-dependent noise measurements in scanning force microscopy, *J. Phys.: Condens. Matter* 8: 7561.

Sader, J. E. (1999). Calibration of rectangular atomic force microscope cantilevers, *Rev. Sci. Instrum.* 70: 3967.

Sader, J. E. & Jarvis, S. P. (2004). Accurate formulas for interaction force and energy in frequency modulation force spectroscopy, *Appl. Phys. Lett.* 84: 1801.

Shustett, M., Burg, T. P. & Manalis, S. R. (2006). Measuring Boltzmann's constant with a low-cost atomic force microscope: An undergraduate experiment, *Am. J. Phys.* 74: 873.

Sugimoto, Y., Innami, S., Abe, M., Custance, O. & Morita, S. (2007). Dynamic force spectroscopy using cantilever higher flexural modes, *Appl. Phys. Lett.* 91: 093120.

Sugimoto, Y., Pou, P., Abe, M., Jelinek, P., Pérez, R., Morita, S. & Custance, O. (2007). Chemical identification of individual surface atoms by atomic force microscopy, *Nature* 446: 05530.

Torrence, T. & Compo, G. P. (1998). A practical guide to wavelet analysis, *Bull. Am. Meteorological Soc.* 79: 61.

Vairac, P., Cretin, B. & Kulik, A. J. (2003). Towards dynamical force microscopy using optical probing of thermomechanical noise, *Appl. Phys. Lett.* 83: 3824.

Yacoot, A. & Koenders, L. (2008). Aspects of scanning force microscope probes and their effects on dimensional measurement, *J. Phys. D: Appl. Phys.* 41: 103001.

Yamanaka, K. & Nakano, S. (1998). Quantitative elasticity evaluation by contact resonance in an atomic force microscope, *Appl. Phys. A* 66: S313.

6

Measurement of the Nanoscale Roughness by Atomic Force Microscopy: Basic Principles and Applications

R.R.L. De Oliveira, D.A.C. Albuquerque, T.G.S. Cruz,
F.M. Yamaji and F.L. Leite[1]
Federal University of São Carlos, Campus Sorocaba
Brazil

1. Introduction

Nanoscale science is the study of objects and phenomena at a very small scale and it has been an emerging, interdisciplinary science involving Physics, Biology, Chemistry, Engineering, Material Science, Computer Science and other areas. The main interest in studying in the nanoscale is related to how nanosized particles have different properties than large particles of the same substance. Nanoscale science allow us to learn more about the nature of matter, develop new theories, discover new questions and answers in many areas including health care, energy and technology and also discover how to make new technologies and products that can improve people's life. Although nanoscale science is a recent development in the scientific community, the development of its main concepts happened over a long period of time and the emergence of nanotechnology is related to experimental advances such as the invention of the Scanning Probe Microscope (SPM), a branch of microscopy that captures surface imagery using physical probes that scan the specimen.

SPM was founded with the invention of the Scanning Tunneling Microscope (STM) in 1982 at IBM in Zurich by Binning (Binning et al., 1982). The tip-sample interaction in STM is based on a tunneling electrical current. Although the ability of the STM to image and measure the material surface with atomic resolution has caused a great impact on the technology community, the tip-sample interaction in STM is limited only for good electrical conductor or semiconductor materials. The need of studying other materials led to the development, in 1986, of the Atomic Force Microscopy (AFM) by Binning, Quate, and Gerber (Binnig et al., 1986) that enabled the detection of atomic scale features on a wide range of insulating surfaces.

2. Basic principles

SPM is defined as a specific type of microscopy that uses the basic principle of scanning a surface with a very sharp probe to image and measure properties of material, chemical and

---

[1] Corresponding Author

biological surfaces. According to the tip-sample interaction the microscopy has a specific name. The two primary forms of SPM are STM and AFM.

The AFM provides a 3D profile on a nanoscale, by measuring forces between a sharp probe (radius less than 10nm) and surface at very short distance (0.2-10nm probe-sample separation). The probe is supported on a flexible cantilever and the AFM tip gently touches the surface and records the small force between the probe and the surface (Wilson & Bullen, 2007). This force can be described using Hooke`s law:

$$F = -k.x \qquad (1)$$

Where F = Force; k = Spring constant; x = Cantilever deflection.

The basic components of an AFM are the probe, the cantilever, the scanner, the laser, a data processor and a photodetector as shown in figure 1.

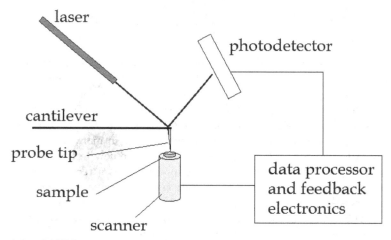

Fig. 1. Principle of AFM.

Forces involved in the tip-sample interaction affect how the probe interacts with the sample. If the probe experiences repulsive forces the probe will be in contact mode otherwise as the probe moves further away from the surface, attractive forces dominate and the probe will be in non-contact mode (Figure 2).

There are three primary imaging modes in AFM: the contact mode where the probe-surface-separation is less than 0.5 nm, the intermittent contact that occurs in a range of 0.5 and 2nm and the non-contact mode where the probe-surface-separation ranges from 0.1 to 10nm.

In contact mode (repulsive regime), if the spring constant of cantilever is less than surface, then the cantilever bends. The force on the tip is repulsive (Figure 3). The forces between the probe and the sample remain constant by maintaining a constant cantilever deflection then an image of the surface is obtained. The advantages of this imaging mode are: fast scanning, good for rough samples and it can be used in friction analysis (Wilson & Bullen, 2007). On the flip side however, forces can damage/deform soft samples (this can be solved by imaging in liquids).

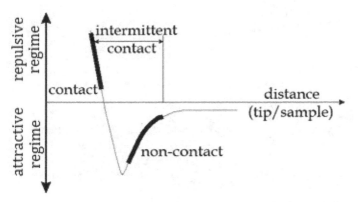

Fig. 2. Force as function of the tip/sample distance. Imaging modes of the AFM based on the type of interaction tip/sample.

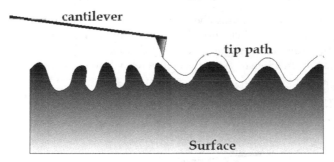

Fig. 3. Contact Mode.

The intermittent mode (Figure 4) is similar to contact but in this mode the cantilever makes intermittent contact with the surface in a resonant frequency (hundreds of KHz). The probe slightly "taps" on the sample surface during scanning, contacting the surface at the bottom of its swing. Because the contact time is a small fraction of its oscillation period, the lateral forces are reduced dramatically. Intermittent mode is usually preferred to image samples with structures that are weakly bound to the surface or samples that are soft (polymers, thin films). There are also two other types of image contrast mechanisms in intermittent mode.

- *Amplitude imaging*: It's an image contrast mechanism where the feedback loop adjusts the z – piezo so that the amplitude of the cantilever oscillation remains (nearly) constant. The voltages needed to keep the amplitude constant can be compiled into an (error signal) image, and this imaging can often provide high contrast between features on the surface (Wisendanger, 1994)
- *Phase imaging*: The main characteristic of this mode is that the phase difference between the driven oscillations of the cantilever and the measured oscillations can be attributed to different material properties. For example, the relative amount of phase lag between the freely oscillating cantilever and the detected signal can provide qualitative information about the differences in chemical composition, adhesion, and friction properties.

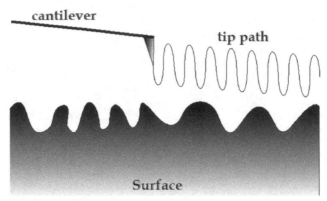

Fig. 4. Intermittent-Contact Mode.

Non-contacting mode (attractive VdW) is based on the knowledge that the probe does not touch the sample but oscillates above it during scanning (Figure 5).

The majority of the samples, unless the ones that are in a controlled UHV (Ultra High Vacuum) or in a environmental chamber have some liquid adsorbed on the surface so the surface topography can be measured by using a feedback loop to monitor changes in the amplitude due to the attractive forces between the probe and the sample.

The advantages of this mode are: very low forces exerted on the samples (10^{-12} N) and extend probe lifetime. The disadvantages of this mode are: generally lower resolution; contaminant layer on surface can interfere with oscillation; usually need UHV to have best imaging.

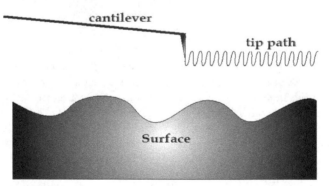

Fig. 5. Non-Contact Mode.

The choice as to which AFM mode to use depends on the surface characteristics of interest and on the hardness/stickiness of the sample. Contact mode is most useful for hard surfaces; a tip in contact with a surface, however, is subject to contamination from removable material on the surface.

Excessive force in contact mode can also damage the surface or blunt the probe tip. Intermittent mode is well-suited for imaging soft biological specimen and for samples with poor surface adhesion (DNA and carbon nanotubes). Non-contact mode is another useful mode for imaging soft surfaces, but its sensitivity to external vibrations and the inherent water layer on samples in ambient conditions often causes problems in the engagement and retraction of the tip.

3. Surface texture: Roughness, waviness and spacing

Surface texture is an important issue when the main interest is to understand the nature of material surfaces and it plays an important role in the functional performance of many engineering components.

The American National Standards Institute's B46.1 specification defines surface texture as the repetitive or random deviation from the normal surface that forms the three dimensional topography of a surface. Before 1990's the measurement of sample surface was obtained by a contact stylus profiler (Whitehouse et al., 1975) that had limitations including a large stylus radius, a large force and low magnification in the plane and may have misrepresented the real surface topography owing to the finite dimension of the stylus tip (Vorburguer & Raja, 1990). On the ultramicroscopic scale of surface, atomic force microscopy (AFM) has been developed to obtain a three-dimensional image of a material surface on a molecular scale.

"Lay" is the term used to indicate the direction of the dominant pattern of texture on the surface. On a surface, the lay is in the front-to-back direction (Figure 6).

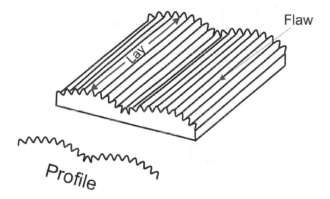

Fig. 6. Surface lay (adapted from B. C. MacDonald & Co., 2011).

Waviness (Figure 7) is the measure of the more widely spaced component of surface texture. It is a broader view of roughness because it is more strictly defined as the irregularities whose spacing, defined as the average spacing between waviness peaks, is greater than the roughness sampling length (Oberg et al., 2000).

There are many parameters for measuring waviness. One of the most important is the waviness evaluation length, which is the length in which the waviness parameters are determined. Within this length the waviness profile is determined. This is a surface texture profile that has the shorter roughness characteristics filtered out, or removed; it also does not include any profile changes due to changes in workpiece geometry. So when it comes to waviness it's important to understand that it's always related to roughness. From this profile the waviness spacing, the average spacing between waviness peaks, is determined.

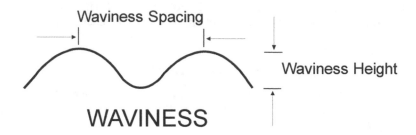

Fig. 7. Waviness (adapted from B. C. MacDonald & Co., 2011).

The waviness height is also determined from the profile, which is just the height from the top of the peak to the bottom of the trough. It is usually at least three times the roughness average height.

Roughness is often described as closely spaced irregularities or with terms such as 'uneven', 'irregular', 'coarse in texture', 'broken by prominences', and other similar ones (Thomas, 1999) (Figure 8). Similar to some surface properties such as hardness, the value of surface roughness depends on the scale of measurement. In addition, the concept of roughness has statistical implications as it takes into consideration factors such as sample size and sampling interval. It is quantified by the vertical spacing of a real surface from its ideal form. If these spacing are large, the surface is rough; if they are small the surface is smooth.

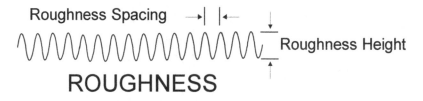

Fig. 8. Roughness (adapted from B. C. MacDonald & Co., 2011).

4. Basic components of AFM

This topic will present some basic ideas about basic components of AFM.

4.1 Scanner

The movement of the tip or sample in the x, y, and z-directions is controlled by a piezo-electric tube scanner, similar to those used in STM. For typical AFM scanners, the maximum ranges are 80 μm x 80 μm in the x-y plane and 5 μm for the z-direction (Figure 9). The scanner moves across the first line of the scan, and back. It then steps in the perpendicular direction to the second scan line, moves across it and back, then to the third line, and so forth. The path differs from a traditional raster pattern in that the alternating lines of data are not taken in opposite directions (Odom, 2004).

While the scanner is moving across a scan line, the image data are sampled digitally at equally spaced intervals. The data are the height of the scanner in z for constant-force mode (AFM) and the data are the cantilever deflection. The spacing between the data points is called the step size. The step size is determined by the full scan size and the number of data points per line. In a typical SPM, scan sizes run from tens of angstroms to over 100 microns and from 64 to 512 data points per line (some systems offer 1024 data points per line.) The number of lines in a data set usually equals the number of points per line. Thus, the ideal data set is comprised of a dense, square grid of measurements.

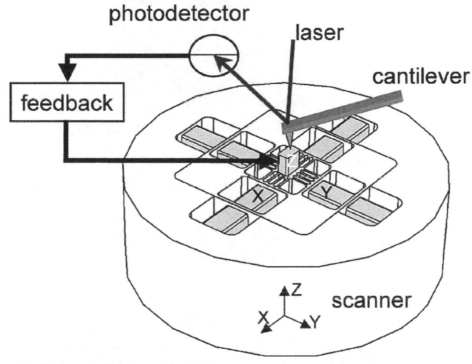

Fig. 9. AFM Scanner (Schitter et al., 2007).

4.2 Probes

The probe is a very important component of a SPM because different probes can measure different properties of the sample (Figure 10). In addition, the probe determines the force applied to the sample. Regarding AFM, the most common probes are the cantilevers that are highly suited to measure the topography of a sample. Different coatings on the cantilevers measure different properties of the sample.

The tip and the cantilever as an integrated component can be fabricated from silicon or silicon Nitride using photolithographic techniques. From a single silicon wafer it is possible to make more than 1000 probes. Regarding the physical properties, the cantilever ranges from 100 to 200 micrometers in length, 10 to 40 micrometers in width, and 0.3 to 2 micrometers in thickness.

Fig. 10. Probe (1nm radius of curvature HI'RES-W probe – MikroMasch, 2011).

4.3 Properties of cantilever

The spring constant of a cantilever (Figure 11) has a critical importance in atomic force microscopy and it is lower than the spring constant between atoms in a solid, which are on the order of 10 N/m. The spring constant of the cantilever depends on its shape, its dimension and the material from which it's made of. Shorter and thicker cantilevers tend to be stiffer and consequently have higher resonant frequencies. Commercial available cantilevers range over four orders of magnitude, from thousands of a Newton per meter to tens of Newton's per meter. Resonant frequencies range from a few Kilohertz to hundreds of Kilohertz allowing high-speed response for non-contact operation.

Fig. 11. Microcantilever with microfabricated tip for a contact-mode AFM. This silicon nitride cantilever was manufactured by Park Scientific Instruments, Mountain View, California. (Photograph by Greg Kelderman – Rugar & Hansma, 1990).

4.4 How to select a probe

The desirable properties for a probe are related to the imaging mode and the application. In contact mode, soft cantilevers are preferable because they deflect without deforming the surface of the sample, a silicon nitride microlever would be a good choice for most applications. In non-contact mode, stiff cantilevers with high resonant frequencies give optimal results. For   applications other than topography, like MFM (Magnetic Force Microscopy), NSOM (Near-field scanning optical microscopy), SThM (Scanning thermal microscopy) etc., different types of probes have to be used. The probes are available mounted on AutoProbe mounts, mounted on TopoMetrix mounts or unmounted in pre-separated quantities or as a half wafer (Howland & Benatar, 2000).

5. Measurement of surface profile

The characterization of surface topography and its understanding is important in procedures involving friction, greasing and wear (Thomas, 1999). Surface measurement determines surface topography, which is essential for conforming a surface's suitability for a specific function. Surface measurement generally includes surface shape, surface finish and surface roughness. For example, engine parts may be exposed to lubricants to prevent potential wear, and these surfaces require precise engineering— at a microscopic level— to guarantee that the surface roughness holds enough of the lubricants between the parts under compression, so as not to make metal to metal contact. For manufacturing and design purposes, measurement is critical to ensure that the finished material meets the design specification. A profilometer is used to measure surface profile as the surface is moved relative to the contact profilometer's stylus, this notion is changing along with the emergence of numerous non-contact profilometery techniques.

A diamond – sharp stylus is used for the measurement of the surface. The pen is placed on an irregular surface at a constant speed for the variation of surface height with horizontal displacement. According to international standards, a pen may have an angle of 60 and 90 degrees and a tip radius of curvature of 2 microns, 5 or 10. A truncated pyramid is one type with a 90 degree included angle between the opposite sides. It is likely that a profile containing many peaks and valleys with a radius of curvature of 10 microns or less and slopes greater than 45 degrees would be misunderstood by such a stylus (Vorburguer & Raja, 1990).

A very important thing to consider is that the variation of the radius of stylus tip may affect the shape of the profiled surface because the radius of the tip of the pen draws a single envelope of the actual profile. The resolution depends on the real contact between the pen and the actual profile. As the radius stylus is increased contact is made with fewer points on the surface, and therefore the profile is modified. Increasing the radius stylus tends to reduce the measured amplitude of the parameter like Ra (Roughness average). However, the relative effect on roughness is not as great as the peak to valley, and other parameters that are best suited for analysis of sensitive surface structures.

The stylus instruments can be used with two different attachments. The first one has a fixed reference, which limits the movement of the pick-up to a horizontal and the transducer gives the height difference between the instantaneous movement of the pen and the whole pick-up. This is the ideal way to measure the surface profile. Unfortunately, this method requires a setup procedure for leveling by a skilled operator. In order to reduce the setup process, a skid can be used as second attachment. The skid is one foot blunt that has a large radius of curvature, and it's placed either beyond or behind the stylus. The transducers sense the difference in level between the stylus and the skid. The skid acts as a mechanical filter to attenuate the long spatial wavelength of the surface. As a result of slippage, the wavelength information is long lost. If the long wavelengths are functionally relevant, then the use of a slide should be avoided. In addition, the use of a skid surfaces or surfaces with periodic discrete peaks can result in distortion of the measured profile (Vorburguer & Raja, 1990).

The numerical evaluation of roughness is always preceded by removal of waviness from the measured profile. This is achieved in a surface texture measuring instrument by using an analog or digital filter. In order to exclude waviness, a limiting wavelength has to be specified. This limiting wavelength is referred to as cutoff. The cutoff is given in mm or Inch and the following values are available in many instruments, 0.08, 0.25, 0.8, 2.5, and 8 nm. The cutoff selected must be short enough to exclude irrelevant long wavelength and at the same time long enough to ensure that enough texture has been included in the assessment to give meaningful results. Usually five cutoff settings are used for assessment, and overall traverse length is seven cutoffs (Vorburguer & Raja, 1990).

6. Traditional surface texture parameters and functions (parameters R and S)

The roughness can be characterized by several parameters and functions (such as height parameters, wavelength parameters, spacing and hybrid parameters (Gadelmawla et al., 2002). The following parameters and functions related to the height and spacing (also called parameters R and S) will be discussed as well as their calculation.

6.1 Height parameters (R)

The most significant parameters in the case of roughness are the Height Parameters.

6.1.1 Roughness average

Among Height Parameters, the roughness average (R_a) is the most widely used because it is a simple parameter to obtain when compared to others. The roughness average is described as follows (Park, 2011).

$$R_a = \frac{1}{L}\int_0^L |Z(x)|\,dx \qquad (2)$$

Where $Z(x)$ is the function that describes the surface profile analyzed in terms of height (Z) and position (x) of the sample over the evaluation length "L" (Figure 12).

Thus, the R_a is the arithmetic mean of the absolute values of the height of the surface profile $Z(x)$. Many times the roughness average is called the Arithmetic Average (AA), Center Line Average (CLA) or Arithmetical Mean Deviation of the Profile. The average roughness has advantages and disadvantages. The advantages include: ease of obtaining the same average roughness of less sophisticated instruments, for example, a profilometer can provide (R_a); possibility of repetition of the parameter, since it appears very stable statistically, recommended as a parameter for the characterization of random surfaces, it is usually used to describe   machined surfaces (B.C. MacDonald & Co, 2011).

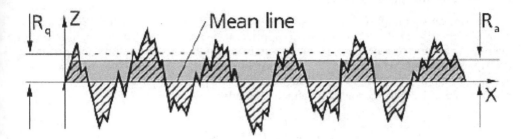

Fig. 12. Profile of a surface (Z). It represents the average roughness R_a and R_q is the RMS roughness based on the mean line (B.C. MacDonald & Co, 2011).

The average roughness, as already said, is just the mean absolute profile, making no distinction between peaks and valleys. Thus it becomes a disadvantage to characterize an average surface roughness if these data are relevant.

The average roughness can be the same for surfaces with roughness profile totally different because it depends only on the average profile of heights. Surfaces that have different undulations are not distinguished (Figure 13). We may have an even surface and some other with peaks (or valleys) with small contributions presenting the same value of average roughness. For this reason, more sophisticated parameters can be used to fully characterize a surface when more significant information is necessary, for example, distinguish between peaks and valleys.

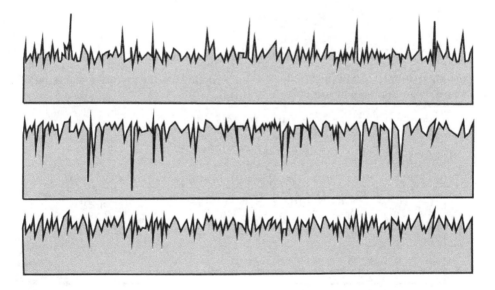

Fig. 13. Different profiles of surfaces, with the same roughness average (adapted from Predev, 2011).

6.1.2 Root mean square roughness (R_q)

The root mean square (RMS) is a statistical measure used in different fields. We cite, as an example, the use of the RMS amplitude applied to harmonic oscillators, such as on an alternating electric current. The root mean square of roughness (R_q) is a function that takes the square of the measures. The RMS roughness of a surface is similar to the roughness average, with the only difference being the mean squared absolute values of surface roughness profile. The function R_q is defined as (Gadelmawla et al., 2002):

$$R_q = \sqrt{\frac{1}{L}\int_0^L |Z^2(x)|\, dx} \tag{3}$$

The software does not need to be sophisticated in order to obtain R_q. For this reason much of the surface analysis equipment (profilometer and SPMs) provides R_q. In SPM, the R_q depends on the swept area of the sample, the scan size.

The R_q is more sensitive to peaks and valleys than the average roughness due to the squaring of the amplitude in its calculation.

6.1.3 Maximum height of the profile (R_T), maximum profile valley depth (R_v) and maximum profile peak height (R_p)

The Maximum Profile Peak Height (R_p) is the measure of the highest peak around the surface profile from the baseline. Likewise the Maximum Profile Valley Depth (R_v) is the measure of the deepest valley across the surface profile analyzed from the baseline (Park, 2011). We can write:

$$R_p = |\max Z(x)| \text{ for } 0 \le x \le L \tag{4}$$

$$R_v = |\min Z(x)| \text{ for } 0 \le x \le L \tag{5}$$

Thus the Maximum Height of the Profile (R_T) can be defined as the vertical distance between the deepest valley and highest peak.

$$R_T = R_p + R_v \tag{6}$$

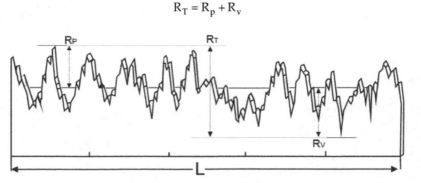

Fig. 14. Illustration of Maximum Height of the Profile (R_T), Maximum Profile Valley Depth (R_v) and Maximum Profile Peak Height (R_p) for the surface profile (adapted from Predev, 2011).

These parameters are useful when trying to find some very sharp peak, which could affect any application of the sample, a scratch or an unusual crack on the material. Often we use the average of these parameters, R_{Tm}, R_{pm} and R_{vm} and comprising the averages along the profile (x) on an evaluation length (L), given by (Park, 2011):

$$R_{pm} = \frac{1}{L}\sum_{i=x}^{L} R_{pi} \tag{7}$$

$$R_{vm} = \frac{1}{L}\sum_{i=x}^{L} R_{vi} \tag{8}$$

$$R_{Tm} = \frac{1}{L}\sum_{i=x}^{L} R_{Ti} = R_{pm} + R_{vm} \tag{9}$$

These average parameters have the same advantage that the extreme parameters described in (4), (5) and (6), but lose accuracy when searching for singularities.

6.1.4 Ten point average roughness - R_z (ISO)

The R_z (ISO) is the arithmetic mean of the five highest peaks added to the five deepest valleys over the evaluation length measured. It is a parameter similar to R_{Tm}.

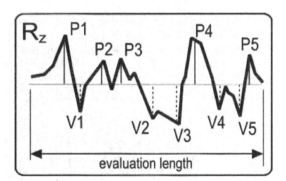

Fig. 15. R_z average of the sum of the five highest peaks in the five deepest valleys of sample's profile (Zygo Corporation, 2011).

6.1.5 Third highest peak to third lowest valley height (R_{3zi})

The Third Highest Peak to Third Lowest Valley Height is a parameter that overlooks the two highest peaks and the two deepest valleys, taking as measure the vertical variation of the third highest peak from the third deepest valley. The advantage of using this parameter is that it disregards any discrepancies that do not interfere in a meaningful way in the characterization of the surface profile. This ignores the extreme values thus reducing the instability of peaks and valleys. It is a commonly used parameter for the characterization of porous surfaces and sealing surfaces (B.C. Macdonald & Co, 2011).

6.2 Roughness spacing parameters

The Roughness Spacing Parameters are the parameters that relate the roughness to the profile of curling and repetition over a surface.

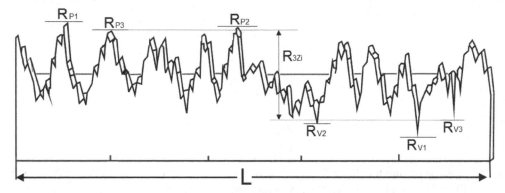

Fig. 16. The two highest peaks (R_{P1} and R_{P2}) and the two deepest valleys (R_{V1} and R_{V2}) are disregarded. The $R_{3z}i$ is counted from the third highest peak (R_{P3}) and the third deepest valley (R_{V3}) (adapted from Predev, 2011).

6.2.1 Peak count

The Peak Count (P_c) is a parameter that provides the count of peaks analyzed along the length L of a surface profile. In this case, the computed peak is the "peak" crossing above an upper threshold and then below the lower threshold. Therefore only values of extreme peaks are significant and establishes a bandwidth in which only peaks and valleys beyond this range will be computed (Park, 2011). The Peak Count is expressed in peaks/inch or peaks/cm (Zygo Corporation, 2011).

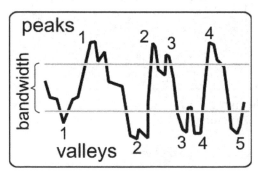

Fig. 17. To determine the peak count only the peaks and valleys that exceed the bandwidth and return are considered (Zygo Corporation, 2011).

6.2.2 Peak density

Peak Density (P_D) represents the density of peaks, i.e., the number of peaks per unit area (Zygo Corporation, 2011).

6.2.3 High spot count

The High Spot Count (HSC) is a parameter similar to the peak count. The main difference between these two parameters is in the defined peak. To be considered a peak in determining the Peak Count, the peak must be followed by a valley that crosses the entire band width (upper and lower threshold). For the High Spot Count, a threshold is set above the average roughness and only the peaks that exceed this one threshold are considered.

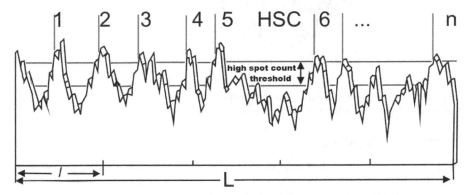

Fig. 18. Illustration of the High Spot Counting and the threshold determining a peak. (Adapted from Park, 2011)

The use of parameters such as Peak Count, Peak Density and High Spot Count has its main application in sheet metal production where the quality control of coatings and paint surfaces is of fundamental importance (B.C. Macdonald & Co, 2011).

6.2.4 Mean spacing

The Mean Spacing (S_m) is the average spacing between peaks in the length of evaluation. In this case the peak is defined as the highest point, along the profile, between a line crossing over the midline and returning below the midline. The spacing between peaks is the horizontal distance between the points where two peaks cross above the midline (Gadelmawla et al., 2002). Thus the mean spacing (S_m) is defined as the average of spacing individual (S_i):

$$S_m = \frac{1}{L}\sum_{i=1}^{L}S_i \qquad (10)$$

The mean spacing (S_m) is generally described in μm or mm.

6.2.5 Average wavelength

The Average Wavelength (λ_a) is a parameter that relates the spacing between local peaks and valleys weighted by their individual frequencies and amplitudes. Thus, the Average Wavelength is given by:

$$\lambda_a = \frac{R_a}{\Delta_a} \qquad (11)$$

Where R_a is the roughness average and Δ_a the mean slope of profile (Park, 2011).

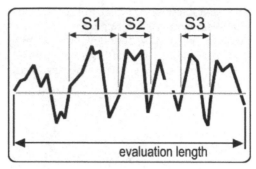

Fig. 19. Three individual spaces in a surface profile. The mean spacing is the average of the three individual spaces on the evaluation length (Zygo Corporation, 2011).

6.2.6 RMS average wavelength

Similar to the R_q and R_a, RMS Average Wavelength (λ_q) takes as reference the root mean square of the spacing between peaks and valleys weighted by their individual frequencies and amplitudes (Park, 2011). RMS Average Wavelength can be calculated by:

$$\lambda_q = 2\pi \frac{R_q}{\Delta_q} \qquad (12)$$

Where R_q is the RMS roughness and Δ_q the RMS slope of profile (Park, 2011).

7. Applications (materials)

The roughness is a very significant parameter for various applications. The characterization of materials through its roughness allows one to obtain information on the efficiency of materials in various application areas.

7.1 Electrochemical intercalation

Some materials, usually transition metals oxides, are classified as intercalation materials. These materials are able to receive short radius ions (H+, Li+...) in its network structure via electrochemical techniques (such as cyclic voltammetry and electrochemical impedance spectroscopy). This process is known as electrochemical intercalation.

The electrochemical intercalation is a reversible process, making these materials very interesting in applications where the control over the ability of insertion/extraction of an ion in a structure is essential.

Intercalation materials are commonly used in micro-batteries, smart windows, smart mirrors, displays, gas sensors and other applications. These materials have potential applications due to the possibility of control of their electronic and optical properties.

The electro-physical-chemical properties of intercalation materials are strongly dependent on surface roughness.

The surface roughness acts as a gateway to allocate the ions in the network structure of the material. In collating materials, height and spacing parameters related to roughness are essential to achieve the efficiency of the material (Cruz et al., 2002).

7.2 Dental enamel

The bleaching process can cause variation in the surface roughness of human tooth enamel. These changes are responsible for color changes, glare reduction, opacity.

The surface roughness is responsible for diffuse scattering of light incident on tooth enamel. Thus the surface roughness is an important variable in the bleaching process and must necessarily be considered.

The search for materials that do not significantly increase the roughness of the tooth surface is a challenge in dentistry. Bleaching agents based on nanoparticles of hidroxiapatatia have been shown to be effective by reducing the surface roughness and increasing the brightness of the tooth enamel (Takikawa et al., 2006).

The brightness and surface roughness are closely associated by an inverse correlation. As shown in the figure 20, the gloss increases with decreasing roughness of tooth enamel (Heintze et al., 2006).

7.3 X-ray anode

The anode is the positive electrode in an X-ray tube. It receives the impact of electrons accelerated by the potential difference due to the high voltage applied. The anode is generally made of materials that have high thermal dissipation, such as copper, molybdenum or rhenium. Depending on the x-ray application (i.e. energy of x-ray) a metal coating such as tungsten (W) or molybdenum (Mo) is placed over these thermally dissipative metals at the impact area of the accelerated electrons.

A change in the spectral distribution of X-rays in a cathode ray tube with increasing roughness of the anode has been observed. The increased surface roughness implies an increase of characteristic peaks and a decrease corresponds to the lower energies of the bremsstrahlung spectrum, and an increase in the average energy of beams of X-rays (Nagel, 1988; Stears et al., 1986). The increased surface roughness implies an increase of characteristic peaks and a decreasing in the part corresponding to the lower energies of the bremsstrahlung spectrum. Unlike the filtration process for tungsten (W) where a dip occurs at lower energies (Yoriyaz et al., 2009).

7.4 Polymeric membranes

In the area of environmental protection, a very significant technology is the process of separation by polymeric membranes. Polymeric membranes, such as Polysulfone / Blend Membrane PLURONIC F127, are used to separate the undesired solute in solution. Thus, the active area of the polymer membrane to carry out the process is the surface. The properties related to the surface are important for performing the separation process. Properties such as the pores size distribution, long-range electrostatic interactions and surface roughness are factors that determine the efficiency of polymer membrane for this application.

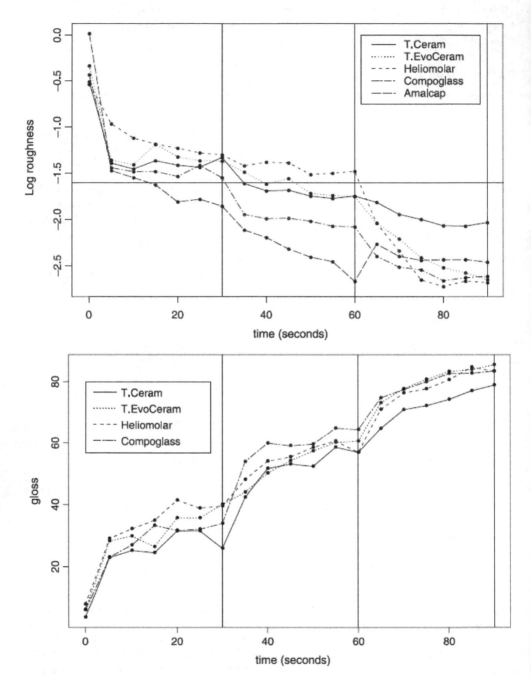

Fig. 20. Relationship between surface roughness and brightness of the teeth treated with five different materials (Heintze & Russon, 2006).

The surface roughness of the polymer membrane is a factor proportional to the bond strength of the membrane. The higher roughness leads to greater adhesive strength of the membrane and greater efficiency in the separation process (Bowen et al., 1998).

8. Effect of RMS roughness on adhesion

In the study of surfaces, related applications to adherence are extremely important. The surface morphology plays a significant effect on adherence. In this section we will discuss the effects that surface roughness plays on adherence.

Adherence is a chemical-physical phenomenon responsible for the union of two surfaces when they come into contact. This union has force of high magnitude in conditions where there is a chemical bond with sharing of electrons, or Coulomb attraction. In some cases the bond strength has relatively low magnitude to occurring by attractive forces of VdW type. The origin of the adhesion force is the same fundamental force of nature responsible for the binding of atoms and molecules.

This phenomenon of interest is multidisciplinary, for example, the effect of adherence on civil engineering projects, cell adhesion in different microorganisms, adhesion of bacteria on the surface of dental enamel, adhesion of polymeric membranes in separation processes for solutes, adhesion of nanoparticles, among others (Bowen et al., 1998).

When contact occurs between two solid bodies, adhesion is not observed. This is due to the fact that much of the surface has a roughness at the microscale. This roughness decreases the area of active interaction between two solid bodies, as only regions with peaks come into contact, thus reducing adhesion.

Liu, D. -L. et al (2007) conducted a study concerning the effect of RMS roughness on the adhesion using AFM. This study provided a better understanding of the effect of roughness on the adhesion when working in the nanoscale. On this scale the effects of adhesion are significant in applications of microelectromechanical systems.

The total adhesion force in this case, the contribution of all molecules involved in the process can be described by the equation (Bowen et al., 1998).

$$F = 2\pi\omega R \left[\frac{R_q}{R + R_q} + \left(\frac{h_c}{h_c + R_q} \right)^2 \right] \tag{13}$$

Where: R = tip radius; R_q = RMS of roughness; h_c = distance separating the tip/sample, and $2\pi\omega R$ represents the strength of the AFM system. The total force is normalized by the surface energy so that ω is the work of adhesion force.

The adhesion force falls with increasing surface roughness and also with increasing radius of the tip used in AFM.

9. Complementary analysis: Fractal dimension and power spectral density

Two powerful techniques for further analysis in the study of surfaces are the fractal dimension and Power Spectral Density. These analysis are based on surface roughness. This type of analysis requires more sophisticated equipment. Most SPMs have image analysis options in their software for additional analysis such as fractal dimension and power spectral density. In this section the analysis will be discussed.

9.1 Fractal dimension

The fractal dimension is a sophisticated parameter used to define the morphology of a surface, considering the roughness present. The surface morphology can be characterized qualitatively by its roughness and its fractal dimension (Guisbiers et al., 2007; Raoufi, 2010; Torkhova & Novikov, 2009; Yadav et al., 2011). The idea of using the concepts of fractal geometry in the study of geometric figures and irregular forms was popularized by Benoit B. Mandelbrot (Mandelbrot, 1982). Since then, such concepts have been used in various fields such as physics, chemistry, biology, materials science, among others.

A fractal is defined by the property of self-similarity or self-affinity, that is, they have the same characteristics for different variations in scale. The thumbnail is like the fractal as a whole and can be classified into self-similar (in the case parts of the fractal is identical to the original fractal) or self-affine (when parts of the fractal is statistically similar to the original). Often fractals are found in nature, for example, when we see the outline of a cloud, the forms produced by lightning, snowflakes, the shape of a cauliflower and especially the morphology of surfaces appear as fractal objects (Figure 21).

Fig. 21. Fractals in nature. (a) Outline of a cloud, (b) lightning, (c) snowflake, (d) surface of a cauliflower, (e) surface of a thin film of nickel oxide obtained AFM (Popsci, 2011; Saint-Marty Marty, 2011; Chaos Theory Dance, 2011).

9.1.1 Self-similar fractals

Self-similar fractals are figures that are completely invariant under scale transformations. An example of self-similar fractal is the Sierpinski Triangle (Figure 22) that does not change its shape under a scale transformation (Assis et al., 2008).

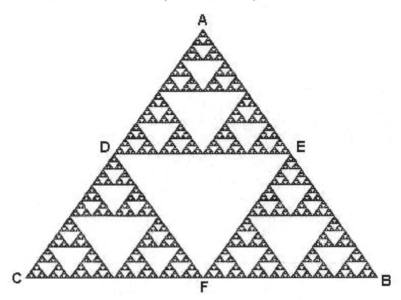

Fig. 22. The Sierpinski Triangle. The Triangle ADE is an exact copy in miniature of triangle ABC, depicting a self-similar fractal object. Triangles CDF and BEF are similarly related to triangle ABC (Assis et al, 2008).

9.1.2 Self-affine fractals

Self-affine fractals are a generalization of self-similar fractals. Self-affine fractal objects are composed of mini-copies of the original figure, but as the scale varies, the proportions are not maintained. Fractals are self-affine fractal invariant under anisotropic transformations. The surfaces of ultrathin films are often treated as self-affine fractal, since during the growth of the films there are two preferred directions of growth (Vicsek, 1989). Likewise most of the surfaces are classified as self-affine fractals.

9.1.3 The concept of dimension

The Euclidean dimension (D), popularly used, is a parameter that defines the geometry of an object. The Euclidean dimension is a parameter in the set of natural numbers in the interval [0, 3]. An object with D = 1, is associated with only one dimension, for example, a line. The dimension D = 2 describes plans objects and dimension D = 3 defines three-dimensional objects. Intuitively, D = 0 describes zero-dimensional objects, as a point, for example.

Not all objects are treated in the field of Euclidean geometry. For many of them can be given a semi-full scale, and this fact characterizes a fractal.

9.1.4 Fractal dimension to self-affine surfaces

Figure 23 represents a two-dimensional self-affine function. This function is defined in the interval [0, 1] and may represent, for example, the profile of a mountain.

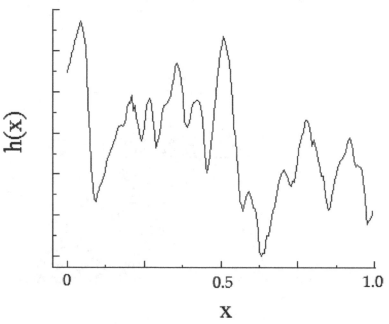

Fig. 23. Hypothetical self-affine function, h (x) represents the height h at position x, with x defined on the interval [0, 1]

Considering the object of study in this chapter, the surface has two preferred directions, one perpendicular to the surface where the roughness characterized by the height parameters, and the other along it. Since the scale is different between the two preferred directions, characterizing the object is an anisotropic transformation (Jurecka et al., 2010). To illustrate the anisotropic transformation, let's imagine that a self-affine surface, described by the function h(x) and expanded by a factor b in the surface dimension (x → bx) then the dimension in the direction of growth must then be magnified by a factor b^α (leading $h \rightarrow b^\alpha h$). This way one can identify the similarities between the original surface and enlarged. Thus preserving the characteristics of an invariant transformation we have (Barabási & Stanley, 1995):

$$h(x) \sim b^{-\alpha} h(bx) \tag{14}$$

Where α, the roughness exponent, provides a quantitative measure of the imperfections of the surface morphology.

Equation (14) provides us with the information that an auto-order of this type is resized horizontally by a factor b, the vertical is just a different factor of b^α. In self-affine fractal objects, the difference between the two points vertically obeys the relation (14), generating (Barabási & Stanley, 1995):

$$\Delta h(x) \sim \Delta x^{\alpha} \qquad (15)$$

Besides the roughness exponent, we can determine other fractal characteristics such as the fractal dimension of self-related functions through methods such as the Box Counting method (Barabási & Stanley, 1995).The profile function h(x) shown in Figure 23 can be embedded within a dimension D = 2. To obtain the fractal dimension through Box Counting method, we can fill this space with objects of equal size to the Euclidian space. Thus, our measuring elements are elements of area with side Δx and area Δx^2. To find the amount of information necessary to cover the function, divide the domain of the function in N_s segments, each of length $\Delta x = 1/N_s$. Making use of equation (15) will require $\Delta h/\Delta x \sim \Delta x^{\alpha-1}$ to cover the function. Finally, the total number of measurement objects to cover the entire function will be:

$$N(\Delta x) \sim N_s . \Delta x^{\alpha-1} \qquad (16)$$

Considering $\Delta x = 1/N_s$ thus:

$$N(\Delta x) \sim \Delta x^{\alpha-2} \qquad (17)$$

From the definition of Fractal Dimension we have (Cruz et al., 2002):

$$D_f = \lim_{\Delta x \to 0}\left(\frac{\ln N(\Delta x)}{\ln(1 / \Delta x)} \right) \qquad (18)$$

The equation (18) for an object of dimension D becomes:

$$D_f = D - \alpha \qquad (19)$$

Equation (19) shows that the surface roughness is closely linked with the fractal dimension. Thus, by knowing the surface roughness exponent one can obtain the fractal dimension.

9.1.5 Fractal dimensions using SPMs techniques

The techniques of scanning probe microscope allow us to find the roughness of surfaces. Figure 24 represents particles striking a one-dimensional surface. Each square represents one particle. The size of the surface in the horizontal direction is defined by an L number of squares.

Using a mathematical tool and the Root Mean Square (RMS roughness) we can find the surface roughness (R_q) using the concepts of Laws Scale (Cruz et al., 2002; Lagally, 1990). In this case, the roughness follows the discrete growth model, shown in Figure 24 as a function of the time and the RMS function is rewritten as (Barabási & Stanley, 1995):

$$R_q^2(L,t) \equiv \frac{1}{L}\sum_{x=1}^{L}\langle[h(x,t) - \langle h_L(x,t)\rangle]^2\rangle_x \qquad (20)$$

Where h(x, t) is the height of the column at x at time t. The term $<h_L(x, t)>$ is the average height of a given observation window L, whereas the brackets $<...>_x$ encompass the spatial

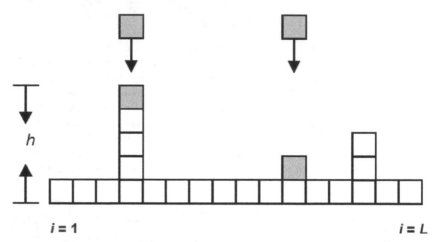

Fig. 24. Diagram illustrating the formation of a surface. Each square represents a particle arriving at this surface (Cruz T.G.S., 2002).

average in x, for all windows of the same size, and averaged over all windows chosen. Using the concepts of fractals, the self-affine function h(x) in this case can be compared to roughness function R(L). Thus, analogously to equation (15) we have:

$$R(L) \sim L^{\alpha} \tag{21}$$

Using equation 20 and plotting a graph of the logarithm of the values obtained from microscopic observations as a function of L we can obtain the roughness exponent α.

Figure 25 is an example of a graph of R_q versus L. For self-affine surfaces, the slope of the roughness in the region of non-saturated is the roughness exponent.

Thus, the roughness exponent can be found and provide the fractal dimension by equation (17).

9.2 Power spectral density (PSD)

The power spectral density (PSD) is a complementary analysis of surface roughness which gives information related to parameters of the roughness height and spacing. The PSD is a parameter used in micrographs which relates the Fourier Transform (FT) with the root mean square roughness (RMS). The relationship between the PSD, FT and RMS is described by (Park, 2011):

$$PSD = FT^2 = RMS^2 \tag{22}$$

A Discrete Fourier Transform (DFT) used for a 3D topography can be described by (Czifra & Horvath, 2011):

$$F(p_k, q_l) = \Delta y . \Delta x \sum_{d=1}^{N} \sum_{c=1}^{M} z(x_c, y_d) e^{-i2\pi(x_c p_k + y_d q_l)} \tag{23}$$

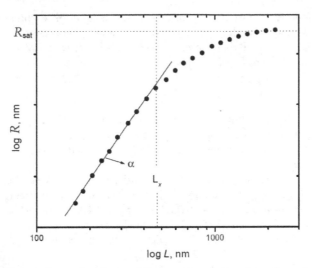

Fig. 25. For self-affine surfaces, the slope of the unsaturated region of roughness is the roughness exponent (Cruz T.G.S., 2002).

Where p_k is the kth frequency in direction x; q_l is the lth frequency in direction y, $z(x_c, y_d)$ is the height coordinate located at x_c, y_d, M is the number of points in the profile, N is the number of profiles, Δx, Δy, are sampling distances, i is the imaginary unit (Czifra & Horvath, 2011).

The PSD is able to separate the surface profile at various wavelengths, that is, the sum of the wavelengths generated by the PSD, make up the original form of the surface profile by a Fourier analysis (Figure 26). This information is relevant in discerning how each component contributes to the original surface.

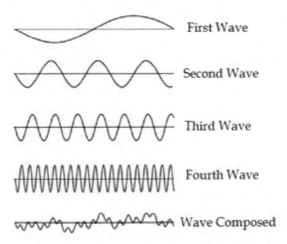

Fig. 26. Example of a wave decomposed by PSD. The sum of the four waves results in the composed wave (adapted from Freitas, A. C. P., 2010).

In the example given in Figure 26, the composed wave equation is the sum of each of the four waves. The surface can thus be divided into height parameters and parameters of wavelength. The figure shows that the first wave contributes to the larger waves while the fourth wave contributes to reduced ripples

10. References

Analytical Sciences Digital Library - Wilson, R. A., & Bullen, H. A. (2007). Introduction to Scanning Probe Microscopy. Department of Chemistry, Northern Kentucky University, Highland Heights, KY 41099. Accessed 10;03;2011. Available from <http://asdlib.org/>.

Assis, T. A., Miranda, J. G. V., Mota, F. B., Andrade, R. F. S., & Castilho, C. M. C. (2008). Geometria fractal: propriedades e características de fractais ideais. *Revista Brasileira de Ensino de Física*, Vol. 30, No 2. (jul 2008) pp. 2304-2314, 1806-1117.

B.C. MacDonald & Co. Accessed 10; 03; 2011. Available from <http://www.bcmac.com/>.

Barabási, A. L., & Stanley, H. E. (1995). *Fractal Concepts in Surface Growth*, Cambridge University Press, 0-521-48318-2, Cambridge

Binnig, G., Quate, C. F., & Gerber, C. (1986). Atomic Force Microscope. *Physical Review Letters*, Vol. 56, No 9, (mar. 1986) pp. 930-933, 0031-9007.

Binning, G., Rohrer, H., Gerber, C., Weibel, E. Surface Studies by Scanning Tunneling Microscopy. *Physical Review Letters*, Vol 49, No 1, pp. 57-61, 0031-9007.

Bowen, W.R., Hilal, N., Lovitt, R.W., & Wright, C.J. (1998). A new technique for membrane characterization: Direct measurement of the force of adhesion of a single particle using an atomic force microscope. *Journal of Membrane Science*, Vol. 139, No 2, (feb. 1998), pp. 269-274, 0376-7388.

Chaos Theory Dance. Accessed 10;03;2011. Available from <http://www.chaostheorydance.com/>.

Cruz, T.G.S. (2002) Leis de Escala e Dimensão Fractal em Filmes Finos: Microscopia de Força Atômica e Técnicas Eletroquímicas. 131 p. Thesis (PhD in physics) – State University of Campinas, UNICAMP, Brazil, Campinas, sep 2002.

Cruz, T.G. S., Kleinke, M.U., & Gorenstein, A. (2002). Evidence of local and global scaling regimes in thin films deposited by sputtering: An atomic force microscopy and electrochemical study. *Applied Physics Letters*, Vol. 81, No 26, (dec. 2002), pp. 4922-4924, 0003-6951.

Czifra, A., & Horvath, S. (2011). Complex microtopography analysis in sliding friction of steel-ferodo material pair. *Meccanica*. Vol. 46, No. 3, (jun 2011), pp. 609-616, 1572-9648.

Freitas, A. C. P. (2010) Influencia do tipo de polimento pós-clareamento na alteração de rugosidade, cor e brilho da superfície de esmalte dental humano. 110 p. Thesis (Phd in Dentistry) – University of São Paulo, USP, Brazil, São Paulo, 2010

Gadelmawla, E.S., Koura, M.M., Maksoud, T.M.A., Elewa, I.M., Soliman, H.H. (2002). Roughness Parameters, *Journal of Materials Processing Technology*, Vol. 123, No 1, (apr. 2002), pp. 133-145, 0924-0135.

Guisbiers, G., Van Overschelde, O., Wautelet, M., Leclere, Ph., & Lazzaroni, R. (2007). Fractal dimension, growth mode and residual stress of metal thin films. *Journal of Physics D-Applied Physics*, Vol. 40, No 4, (feb. 2007), pp. 1077-1079, 0022-3727.

Heintze, S.D., Forjanic, M., & Rousson, V. (2006). Surface roughness and gloss of dental material as a function of force and polishing time in vitro. *Dent Mater*, Vol. 22, No 2, (feb. 2006), pp. 146-165, 1608-4582.

Howland, R. & Benatar, L., (2000). A Practical Guide to Scanning Probe Microscopy, in, *Penn State Polymer Physics Group*, Accessed 10;03;2011, Available from <*http://raman.plmsc.psu.edu/*>.

Jurecka, S., Kobayashi, H., Takahashi, M., Matsumoto, T., Jureckova, M., Chovanec, F., & Pincik, E. (2010). On the influence of the surface roughness onto the ultrathin SiO(2)/Si structure properties. *Applied Surface Science*, Vol. 256, No 18, (jul 2010), pp. 5623–5628, 0169-4332.

Lagally, M. G. (1990). *Kinetics of Ordering and Growth at Surfaces*, Plenum Press, 978-0306437021, New York.

Liu, D.-L., Martin, J., & Burnham, N.A. (2007). Optimal roughness for minimal adhesion. *Applied Physics Letters*, Vol. 91, No. 4, (Jul. 2007), pp. 31071-31073, 0003-6951.

MikroMasch. Accessed 10; 03; 2011. Available from <*http://www.spmtips.com/*>.

Mandelbrot, B. B. (1982). *Fractal Geometry of Nature*, Freeman, 978-0716711865, San Fracisco.

Nagel, H. D. (1988). Limitations in the determination of total filtration of X-ray tube assemblies. *Physics in Medicine and Biology*, Vol. 33, No 2, (feb. 1988), pp. 271-289, 0031-9155.

Oberg, E., Jones, F. D., Horton, H. L., & Ryffel, H. H. (2000). *Machinery's handbook* (26th ed.), Industrial Press Inc., 978-1-59124-118-8, New York.

Odom, T. W., (2004). Laboratory manual, Nanoscale pattern and systems, in, *The Odom Group*, Accessed 10;03;2011, Available from <*http://chemgroups.northwestern.edu/odom/*>.

Park Systems. Accessed 10; 03; 2011. Available from <*http://www.parkafm.co.kr/*>.

Popular Science. Accessed 10; 03; 2011. Available from <*http://www.popsci.com*>.

Precision Devices, Inc. Accessed 10;03;2011. Available from <*http://www.predev.com/*>.

Raoufi, D. (2010). Fractal analyses of ITO thin films: A study based on power spectral density, *Physica B*, Vol. 405, No 1, (jan 2010), pp. 451–455, 0921-4526.

Rugar, D.,Hansma, P. (1990). Atomic Force Microscopy, *Physics Today*, Vol. 43, No 10, (oct. 1990), pp. 23-30, 0031-9228.

Sainty Marty-Marty. Accessed 10;03;2011. Available from <*http://saintmarty-marty.blogspot.com/*>.

Schitter, G., Astrom, K.J., DeMartini, B.E., Thurner, P.J., Turner, K.L., Hansma, P.K. (2007). Design and Modeling of a high-speed AFM-scanner . *IEEE Transacions on Control Systems Technology*, Vol. 15, No 5, (sep. 2007), pp. 906-915, 1063-6536.

Stears, J. G., Felmlee, J. P., & Gray, J. E. (1986). Half-Value-Layer increase owing to tungstein buildup in the X-ray tube: fact or fiction. *Radiology*, Vol. 160, No 3, (sep. 1986), pp. 837-838, 3737925.

Vicsek, T. (1989). *Fractal Growth Phenomena* (2nd ed), World Scientific, 981-0206690, New Jersey.

Vorburguer, T. V., & Raja, J., (1990). Surface Finish Metrology Tutorial, in, *Nist Calibrations*, NISTIR 89-4088

Takikawa, R., Fujita, K., Ishizaki, T., & Hayman, R.E. (2006). Restoration of post-bleach enamel gloss using a non-abrasive, nano-hydroxyapatite conditioner, *Proceedings of 84th General Session & Exhibition of the IADR*, ISBN, Brisbane-Australia, June 28-July 1.

Thomas, T.R. (1999). *Rough Surfaces* (2nd ed), Imperial College Press, 978-1-86094-100-9, London.

Torkhov, N. A., & Novikov, V. A. (2009). Fractal geometry of the surface potential in electrochemically deposited platinum and palladium films. *Semiconductors*, Vol. 43, No 8, (aug 2009), pp. 1071–1077, 1063-7826.

Whitehouse, D.J. (1975) "Stylus Techniques", in Characterization of Solid Surfaces, P.E. Kane and G.R. Larrabee, eds. Plenum Press, New York , p. 49.

Wiesendanger, R. (1994). *Scanning Probe Microscopy and Spectroscopy*, Cambridge University Press, 0-521-42847-5, Cambridge.

Yadav, R.B S., Papadimitriou, E.E., Karakostas, V.G., Shanker, D., Rastogi, B.K., Chopra, S., Singh, A. P., & Kumar, S. (2011). The 2007 Talala, Saurashtra, western India earthquake sequence: Tectonic implications and seismicity triggering. *Journal of Asian Earth Sciences*, Vol. 40, No 1, (jan 2011), pp. 303-314, 1367-9120.

Yoriyaz, H., Moralles, M., Siqueira, P.T.D., Guimarães, C.C., Cintra, F.B., & Santos, A. (2009). Physical models, cross sections, and numerical approximations used in MCNP and GEANT4 Monte Carlo codes for photon and electron absorbed fraction calculation. *Medical Physics*, Vol. 36, No 11, (oct. 2009), pp. 5198-5213, 0094-2405.

Zygo Corporation. Accessed 10;03;2011. Available from <*http://www.zygo.com/*>.

7

Nanoscale Effects of Friction, Adhesion and Electrical Conduction in AFM Experiments

Marius Enachescu
University POLITEHNICA of Bucharest*
Romania

1. Introduction

Friction and adhesion are two so related phenomena of the contact formed by two bodies. And due to the presence of friction and adhesion very often we have wear, i.e., third body presence generated by friction and adhesion. Likely, friction is one of the oldest phenomena in the history of humankind and of natural science, e.g., physics. Friction was the origin of first fire lit by human in early Stone Age, and of the many events Egyptians had faced while pulling huge blocks of stone needed for their pyramids. In fact, Egyptians were basically the first tribologists in history, even if the term tribology defined as "science and technology of interacting surfaces in relative motion", was suggested only in 1966 by Peter Jost.

Humanity needed to enter into Renaissance period in order to have Leonardo da Vinci (1452-1519) (Dowson, 1979) to introduce the first modern concepts of friction. Da Vinci came to two important conclusions:

1. *"Friction produces double the amount of effort if the weight be doubled."*
2. *"The friction made by the same weight will be of equal resistance at the beginning of the movement, although the contact may be of different breadths or lengths."*

In other words, these are today the two fundamental laws of friction, the friction force is proportional to the load (normal force), and independent of the apparent area of contact between the sliding body and the surface.

Two centuries later, Amonton (1663-1705) rediscovered and extended da Vinci's observations. Amonton confirmed these observations with further experiments, from which came Amonton's Law of Friction: $F_f = \mu L$, which states that the friction force F_f is proportional to the applied load L. Thus, today the two fundamental laws of friction are called "da Vinci-Amonton's laws".

It took nearly an extra century, 1785, for the experiments of Coulomb (1736-1806) to distinguish between friction during sticking and sliding. He observed that the coefficient of kinetic friction was generally smaller than the coefficient of static friction. He also observed that μ was generally independent of sliding velocity. Investigating the origins of friction, Coulomb suggested that roughness (asperities) on the micrometer scale is responsible for the occurrence of friction, as depicted in Fig 1. However, there was experimental evidence against his hypothesis: highly polished surfaces did not exhibit low, but high friction.

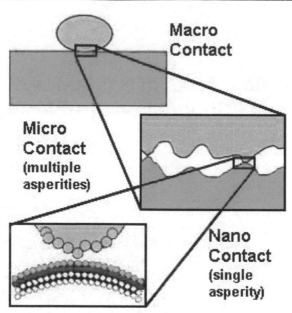

Fig. 1. A macroscopic contact that appears conforming and continuous is usually composed of multiple contact points between many microasperities. The frictional behavior of such a contact follows Amonton's Law. The friction law for a microscopic contact, a single asperity contact, is not known. A scanning probe instrument provides a well-defined single asperity contact (the tip) where interaction forces can be precisely measured with nanometer/atomic resolution. At this scale, macroscopic physical laws no longer apply. For example, the friction force (Ff) is no longer linearly proportional to the applied load (L).

An alternative explanation was given by Desaguliers, who suggested that molecular adhesion might be the relevant phenomenon. However, molecular adhesion was known to be proportional to contact area, whereas friction was found to be independent of contact area.

It is astonishing that wear phenomena, despite their obvious significance, were studied quite late. The reason for this delay may lie in the fact that the leading cause of wear is through the interactions of micro-contacts, which became an object of tribological research only after the work of Bowden and Tabor. It took about two centuries beyond Coulomb's work until this controversy was solved. Around 1950, Bowden and Tabor performed systematic, tribological experiments which showed that the contact of a macroscopic body is formed by a number of small asperities (Fig. 1). Thus, another contact area, the real area of contact had to be introduced. This new concept was extremely successful and is the basics of most present tribological studies. Essentially, the Bowden-Tabor model states that friction is proportional to the real area of contact.

From this point of view, Desaguliers was right to assume that adhesion, which is also proportional to the contact area, is more related to friction than roughness. Therefore, the model is also called Bowden-Tabor adhesion model. In first approximation, the real area of contact does not depend on the apparent contact area. By increasing the load, the number of contacting asperities also increases with load.

The Bowden-Tabor adhesion model explains the da Vinci-Amonton's laws of the macroscopic world. However, a basic understanding of friction is still lacking and many questions remain unanswered such as: i) What are the microscopic mechanisms of friction? ii) How is energy dissipated? iii) How do lubricants (third body presence) affect the shear properties? iv) Can the friction be calculated from molecular interaction potentials in a quantitative way?

During the last two decades, the field of tribology at the atomic and nanometer scale became of interest to a bigger scientific community. These problems are beginning to be addressed by relative recent development of several experimental techniques (Krim, 1996). Instruments such as the surface forces apparatus (SFA) (Israelachvili, 1972) (Israelachvili et al., 1990), the quartz-crystal microbalance (QCM) (Krim et al., 1990) (Watts et al., 1990) (Krim et al., 1991), the atomic force microscope (AFM) (Mate et al., 1987) (Binnig et al., 1986a) and others are extending tribological investigations to atomic length and time scales. Furthermore, advances in computational power and theoretical techniques are now making sophisticated atomistic models and simulations feasible (Harrison & Brenner, 1995). Nanotribology, is the emerging field that attempts to use these techniques to establish an atomic- and nano-scale understanding of interactions between contacting surfaces in relative motion (Carpick et al., 1998a; Enachescu et al., 1998, 1999a, 1999b, 1999c, 2004; Park et al., 2005a; Carpick & Salmeron, 1997; Grierson &. Carpick, 2007; Szlufarska et al., 2008).

Nanotribology, and particularly AFM experiments, focusing on the fundamentals and basic understanding of friction, adhesion, and wear, is trying to do this in terms of chemical bonding and of the elementary processes that are involved in the excitation and dissipation of energy modes.

In AFM experiments, besides the obvious friction and wear obvious experiments, adhesion measurements are easily performed via so called pull-off experiments. A basic pull-off experiment is described in Fig. 2 below.

Regarding the excitation and dissipation of energy modes during tribological experiments, several mechanisms have been investigated and proposed. One is related to coupling to the substrate (and tip) electron density that causes a drag force, similar to that causing an increase of electrical resistance by the presence of surfaces in thin films (Daly & Krim, 1996; Sokoloff, 1995; Persson & Volokitin, 1995; Persson & Nitzan, 1996). Another is related to excitation of surface phonon modes in atomic stick-slip events. Delocalization of the excited phonons by coupling to other phonon modes through nonharmonic effects and transport of the energy away from the excited volume leads to efficient energy dissipation (Sokoloff, 1993; Carpick & Salmeron, 1997). At high applied forces, an important event is the wear process leading to rupture of many atomic bonds, the creation of point defects near the surface, displacement and creation of dislocations and debris particles.

Another level of our understanding focus, and where AFM experiments may decisively contribute, includes questions such as the nature of relative motion between the two contacting bodies: is it continuous (smooth sliding) or discontinuous (stick-slip, e.g., atomic stick-slip)? How does friction depend upon the actual area of contact between two bodies? Are friction and adhesion related, and how? What is the behavior of lubricant molecules, including third bodies, at an interface? How are they compressed and displaced during loading and shear? How does their behavior depend upon their molecular structure and chemical identity? It is our intent to partially address some of these questions in the work presented here.

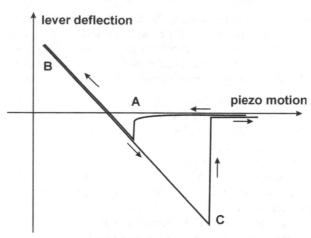

Fig. 2. A typical pull-off, or force vs. displacement curve during an approach-retract experiment. The AFM lever deflection is recorded while tip and sample are brought into contact and separated again. As long as tip and sample are separated, the free lever is not deflected. When tip and sample are brought into close proximity (A) the lever "feels" an increasingly attractive force, caused by long range electrostatic or van der Waals forces, and is bent down towards the sample. When the force gradient exceeds the spring constant of the lever, an instability, the so called "jump-to-contact", occurs, and the tip abruptly contacts the surface. Upon further approach, the lever experiences a repulsive force and is bent upwards (B), for small deflections following Hook's law. Upon reversal of the piezo motion the deflection signal follows the previous path. But adhesive forces will keep the tip in contact with the sample until the elastic force of the lever exceeds the adhesion and the lever snaps out of contact (C).

As mentioned, friction and adhesion are related phenomena of the contact formed by two bodies. And due to the presence of friction and adhesion very often we have wear, i.e., third body presence generated by friction and adhesion. Also, the interface created under friction and adhesion plays a crucial role in the local electrical conductivity between the two bodies, besides the bodies' intrinsic conductivity properties. All of these, friction, adhesion, third body presence (wear) and local conductivity at nanometer scale are the goal of this work, that is trying to bring extra light in the emerging nanotribology field.

2. Atomic- and nano-scale friction experiments on a special interface

The nanotribological properties of a hydrogen-terminated diamond(111)/tungsten-carbide interface have been studied using ultra-high vacuum atomic force microscopy. Both friction and local contact conductance were measured as a function of applied load. The contact conductance experiments provide a direct and independent way of determining the contact area between the conductive tungsten-carbide AFM-tip and the doped diamond sample. It was demonstrated that the friction force is directly proportional to the real area of contact at the nanometer-scale. Furthermore, the relation between the contact area and load for this extremely hard heterocontact is found to be in excellent agreement with the Derjaguin-Müller-Toporov continuum mechanics model.

According to the classical law of friction, the friction force between two bodies in motion is proportional to the applied load and independent of the apparent area of contact (Dowson, 1998). However, a macroscopic contact between two apparently flat solid surfaces consists in practice of a large number of micro-contacts between the asperities that are present on both contacting surfaces (Fig. 1). The classical law of friction, which cannot be understood or deduced from first principles, is the result of many complex phenomena at the interface, in particular the specific interactions between contacting asperities, and the corresponding deformations of these asperities (Greenwood & Williamson, 1966). Although macroscopic tribological research can provide important empirical information about the frictional behavior of materials, it cannot explain friction at a fundamental level. Only detailed studies of friction at a single-asperity contact, under well-defined conditions and with nanometer-scale or even atomic-scale resolution, can result in an understanding of friction at a fundamental level. Some ultra-high vacuum atomic force microscopy (UHV-AFM) experiments indicate that friction is proportional to the contact area for a nanometer-sized single-asperity contact (Carpick et al., 1996a, 1998a; Enachescu et al., 1998; Lantz et al., 1997a, 1997b). In some of these studies, the contact area was not directly measured but instead derived from continuum mechanics models, although, as discussed further below, it is generally not clear *a priori* which model is valid for a specific combination of materials. As well, most of these experiments were performed on layered materials, where it is unclear whether continuum mechanics models can be used quantitatively. Nevertheless, the continuum mechanics models generally provided convincing fits to the data. Carpick *et al.* (Carpick et al., 1996a) performed experiments on muscovite mica and found that friction was proportional to the contact area as described by the Johnson-Kendall-Roberts (JKR) model (Johnson et al., 1971). Experiments by Lantz *et al.* (Lantz et al., 1997a, 1997b) on NbSe$_2$ and graphite resulted in a relation between friction and contact area as described by the Maugis-Dugdale (MD) model (Maugis, 1992; Johnson, 1997). Only one observation of the Derjaguin-Müller-Toporov (DMT) model (Derjaguin et al., 1975; Müller et al., 1983) has been reported so far by Enachescu *et al.* (Enachescu et al., 1998). The experiments were conducted with an extremely hard heterocontact, involving stiff materials with low adhesive forces, *i.e.* a tungsten-carbide AFM-tip in contact with a hydrogen-terminated diamond(111) sample. Both diamond and tungsten-carbide are extremely stiff, non-layered materials. Furthermore, hydrogen passivates the diamond surface while carbides are generally quite inert.

In this study, we discuss the results of a nanotribological study of a hydrogen-terminated diamond(111)/tungsten-carbide single asperity interface using UHV-AFM. Since the diamond sample is slightly boron-doped and the tungsten-carbide tip is conductive, we are able to measure the local contact conductance as a function of applied load. These experiments provide an independent way of determining the contact area, which can be directly compared to the corresponding friction force. Diamond and diamond-like films are important coating materials used in a wide variety of tools, hard disks, micro-machines, and aerospace applications. For micro-machine and hard disk applications in particular, the nanotribological properties are of great importance (Seki et al., 1987). Similarly, tungsten-carbide plays an important role in several types of hard coatings (Schwartz, 1990).

2.1 Background

The AFM results (Carpick et al., 1996a, 1998a; Enachescu et al., 1998; Lantz et al., 1997a, 1997b) and surface forces apparatus (Homola et al., 1989) experiments indicate that the friction force F_f varies with the applied load L in proportion to the tip-sample contact area A. Thus, $F_f = \tau A$ where τ is the shear strength, a fundamental interfacial property. In most cases, the relation between A and L is deduced from elastic continuum mechanics models, assuming a sphere (tip) in contact with a flat plane (sample) (Johnson, 1987). However, the correct relation between A and L not only depends on the exact geometry but also upon the strength of the adhesive forces compared to the elastic deformations (Maugis, 1992; Müller et al., 1980; Tabor, 1977; Greenwood, 1997; Johnson, 1996).

The JKR and DMT models mentioned above have been deduced for two extreme cases, namely for compliant materials with strong, short-range adhesive forces and for stiff materials with small, long-range adhesive forces, respectively. The empirical nondimensional parameter $\mu = \left(R\gamma^2 / E^{*2} z_0^3 \right)^{1/3}$ can be used to determine which of the two continuum mechanics models is most appropriate (Tabor, 1977; Johnson, 1996). In this expression, R is the sphere radius, γ is the work per unit area required to separate tip and surface from contact to infinity, and E^* is a combined elastic modulus, given by the equation $E^* = \left((1 - v_1^2) / E_1 + (1 - v_2^2) / E_2 \right)^{-1}$, where E_1 and E_2 are the Young's moduli, and v_1 and v_2 are the Poisson's ratios of the sphere and plane, respectively. Finally, z_0 represents the equilibrium spacing for the interaction potential of the surfaces. If $\mu > 5$, the JKR theory should be valid, while for $\mu < 0.1$, the DMT theory should describe the relation between A and L (Tabor, 1977; Johnson, 1996; Greenwood, 1997). Neither the JKR nor the DMT limit is appropriate for the intermediate cases ($0.1 < \mu < 5$). As discussed by Greenwood (Greenwood, 1997), it is difficult to calculate the exact area of contact for the continuum problem. Greenwood obtained a numerical solution using a Lennard-Jones potential and defined the contact edge as the point of maximum adhesive stress. Greenwood's solution closely resembles the Maugis-Dugdale model. In both cases, the variation of contact area with load then appears very close to the *shape* of the JKR curve for values of $\mu > 0.5$. However, the JKR equation does not correctly predict the *actual* contact area, pull-off force, and thus the adhesion energy, unless $\mu > 5$. Therefore, while a measurement of contact area versus load may resemble a JKR curve, quantitative analysis would be uncertain, as it would highly depend on a specific model for the tip-sample interaction potential.

In the case of the DMT model ($\mu < 0.1$), the contact area A varies with the applied load L in a simple fashion: $A = \pi \dfrac{R^{2/3}}{K^{2/3}} \left(L + 2\pi\gamma R \right)^{2/3}$, where K=(4/3)E*. The pull-off force or critical load L_c is given by $L_c = -2\pi\gamma R$. The value of L_c can be obtained from AFM approach/retract displacements of the cantilever and sample, by measuring the (negative) normal force required to separate tip and sample. We note that the contact area goes to zero at pull-off, in contrast to the JKR model.

The contact radius in AFM experiments is generally in the nanometer-range and, consequently, much smaller than the electronic mean free path. In this limit, the contact conductance becomes directly proportional to the contact area, as described by Sharvin's equation for metallic contacts (Jansen et al., 1980): $G = 3\pi a^2 / 4\rho l$, where ρ is the resistivity, l is the mean free path of the conduction electrons, and a is the radius of the contact. We stress that this equation is only valid for nanometer-sized contacts, where $l \gg a$. The linear relationship between the contact conductance and contact area is true whether the junction is Ohmic, semiconductor-like, *etc.* For instance, in the case of a metal/semiconductor contact (Sze, 1981), which matches our tip-sample interface, the current is directly proportional to the area of contact, considering a constant metal/semiconductor barrier height and a constant temperature during the experiments. We do not expect to observe the current to change step-wise with load, *i.e.*, the well-known phenomena of quantized conductance occurring at contacts consisting of only a few atoms (Rubio et al., 1996), since in our experiments the contact area contains many atoms.

2.2 Experimental

The experiments were performed in an UHV chamber (base pressure 7 x 10^{-11} Torr), since even in moderately evacuated chambers the residual oxygen and water vapor may combine with the sliding action to catalyze a phase change on diamond (Gardos, 1994). The UHV-chamber is equipped with a home-built AFM (Dai et al., 1995), low-energy electron diffraction (LEED), and Auger electron spectroscopy (AES). The sample is an artificial type IIb diamond(111) single-crystal, which is terminated with hydrogen and slightly boron-doped. The cleaning procedure used, as well as the single-crystal quality, are described in more detail by van den Oetelaar *et al.* (van den Oetelaar & Flipse, 1997). Fig 3 shows the LEED pattern taken after the cleaning procedure. This clear (1x1) LEED pattern supports the fact that we have a hydrogen-terminated diamond(111)-(1x1) surface.

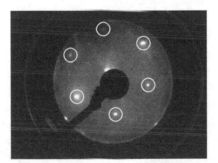

Fig. 3. Our cleaning procedure gave rise to a clear hydrogen-terminated diamond(111)-(1x1) surface, as shown in this LEED pattern.

Triangular silicon cantilevers with integrated tips, coated with approximately 20 nm tungsten-carbide, were used for all measurements. The tips were characterized by scanning electron microscopy (SEM) and AES. Two types of cantilevers were used, with a spring constant of 88 N/m and 0.23 N/m, respectively. The former cantilever was used for conductance measurements while the latter one was used for friction measurements. The tips were cleaned in UHV immediately prior to the measurements, by applying short voltage pulses and/or by rubbing them on the surface. Normal cantilever force constants were taken from the manufacturer, and the normal/lateral force ratio was calculated using the method described by Ogletree *et al.* (Ogletree et al., 1996). The absolute accuracy of the forces measured is limited due to significant uncertainty in the material properties of the cantilever and approximations used in the force constant calculations. However, relative changes in friction could be accurately determined by using the same cantilever and tip during a series of measurements. A flexible I-V converter, allowing current measurements spanning the range from pA to mA, was designed and built.

Friction versus load data were acquired by scanning the AFM-tip repeatedly back and forth over the same line on the surface, while linearly increasing or decreasing the externally applied load. The value of the friction force at a given load is half of the difference between the signals while scanning from left to right, and right to left, respectively (Carpick et al., 1996a; Hu et al., 1995).

2.3 Results and discussions

All of the results presented in this work were obtained on a hydrogen-terminated diamond(111) sample, consisting of atomically smooth and well-ordered islands of 150 - 250 Å in diameter (Enachescu et al., 1998; van den Oetelaar & Flipse, 1997). The friction and contact conductance data were acquired within the bounderies of a single island, thus avoiding multiple-contact points.

Fig. 4(a) shows a large number of I-V curves recorded at different loads up to 1.7 μN, using an 88 N/m cantilever. The I-V characteristics are semiconductor-like and consistent with the p-type doping of the diamond sample. The shape of the I-V curves remains basically constant at all loads, strongly indicating that the applied load does not significantly affect the surface electronic properties of the interface. This observation supports our assumption that the current is proportional to the contact area.

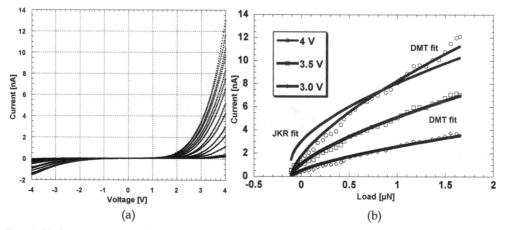

Fig. 4. (a) Current measured through the tip-sample contact versus bias voltage (I-V curves) recorded as a function of increasing load up to 1.7 μN. (b) Current versus applied load at three different constant bias voltages. The DMT fit is significantly better than the JKR fit, as illustrated for a bias voltage of 4 V, also indicated by the mean square deviation of the JKR fit, which is more than one order of magnitude worse for the JKR fit compared to the DMT fit.

Plotted in Fig. 4(b) is the load dependence of the current at several bias voltages applied to the sample, e.g., +3 V, +3.5 V and +4 V. The data can be fit by the DMT model, using Lc as a free parameter. The DMT model provides an excellent fit to the measured data, and the value of L_c deduced from the fits is in excellent agreement with the independently measured pull-off force of -0.1 μN, obtained from cantilever-sample retract experiments for the same cantilever.

The current versus load data was fitted using the JKR model. Treating L_c as a free parameter, the JKR fits, at all bias voltages, predict a critical load which is systematically and substantially too small compared to the independently measured pull-off force. If we apply the constraint $L_c = -0.1$ μN to the JKR fit, the resulting fit is clearly incompatible with our data, as illustrated in Fig. 4(b) for a bias voltage of 4 V. In addition, we found from the fitting statistics that the mean square deviation of the JKR fit is more than one order of magnitude worse than that of the DMT fit. These local contact conductance results clearly show that the load dependence of the contact area for this single-asperity interface can be described by the DMT continuum mechanics model.

A topographic AFM image is actually a convoluted image of the tip and surface features of the sample. Usually, one requires sharp AFM-tips to reveal the surface topography, but similarly, an extremely sharp feature on the surface can provide information about the shape of an AFM-tip (Atamny & Baiker, 1995). To determine the radius of curvature of the tungsten-carbide coated tip used in our friction experiments, we performed scans over the sharp edges of a faceted $SrTiO_3(305)$ sample (Sheiko et al., 1993) in air. The surface is terminated with a large number of (101) and (103) facets, which form long sharp ridges that are suitable for tip imaging (Carpick et al., 1996a; Ogletree et al., 1996; Sheiko et al., 1993). The thus obtained cross-sectional "image" of the AFM-tip actually provides an upper limit to the tip dimensions, but this upper limit appears to be very close to the real tip dimensions (Carpick et al., 1996a). The cross sectional tip-profile can be fit by a hemisphere, as is shown in Fig. 5, resulting in a radius of curvature of 110 nm. Profile analysis using the $SrTiO_3(305)$ sample was performed before and after tip-sample contact, and no evidence of wear was discerned.

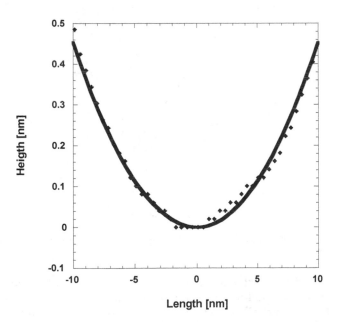

Fig. 5. Hemispherical fit of the AFM-tip profile, resulting in a radius of curvature $R = 110$ nm. Note the difference in vertical versus horizontal scale.

Having obtained a value for the tip radius R, we can estimate the empirical parameter μ. Using $L_c = -2\pi\gamma R$, γ can be obtained from the measured pull-off force. A typical normal force versus cantilever-sample displacement curve, during retraction of the cantilever, is shown in Fig. 6. The corresponding pull-off force is -7.3 nN, resulting in $\gamma = 0.01$ J/m^2. Thus, using $z_0 = 2$ Å, $E_{diamond} = 1164$ GPa (Klein, 1992), $E_{WC} = 700$ GPa (Shackelford et al., 1994), $\nu_{diamond} = 0.08$ (Klein, 1992), and $\nu_{WC} = 0.24$ (Shackelford et al., 1994), we find that $\mu = 0.02$. Indeed, this value is much smaller than the DMT condition $\mu < 0.1$ discussed above, showing that the present tip-sample contact is firmly in the DMT regime.

Friction experiments were performed as a function of applied load using the soft lever (Enachescu et al., 1998). They were reproducible at different locations on the sample, and were obtained by decreasing the load from 12 nN to negative loads (unloading). Experiments where the load was increased (loading) exhibited the same behavior as the unloading results, indicating that the deformation of the contact is elastic for the range of loads investigated.

Friction versus load experiments could be fit very well by the DMT model, while treating both γ and the shear strength τ as free parameters (Enachescu et al., 1998). The mathematical fit results in a pull-off force of -7.3 nN and a shear strength of 238 MPa. Thus, the pull-off force predicted by the DMT fit is in excellent agreement with the pull-off value measured experimentally, as shown in Fig. 6.

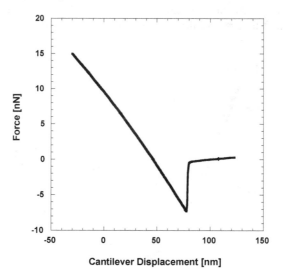

Fig. 6. Typical normal force versus cantilever-sample displacement curve, during retraction of the cantilever. The corresponding pull-off force is –7.3 nN.

The measured pull-off force actually represents an *independent* verification of the DMT fit, since γ (and thus also the pull-off force) was treated as a free parameter in the DMT fit. Attempts to fit the JKR model to the friction versus load curves, using L_C both as a free parameter and as a constrained parameter, produced strongly inconsistent fits. Experimentally, no friction data for loads smaller than -2 nN could be obtained due to a premature pull-off of the tip. This premature pull-off is promoted by the tip-sample movement during scanning and is more likely to appear in this particular experiment due to the very low adhesive force between the surfaces in contact.

In an attempt to learn more about the relation between the friction force and the area of contact, we have plotted the friction force versus contact area, and the result is shown in Fig. 7. The friction force plotted in this figure is exactly the friction force measured during friction versus load experiments. The contact area was calculated using the DMT theory. The use of the DMT theory is supported by the three previous pieces of experimental evidence, namely: (i) the excellent DMT fit of the current versus load data using the stiff lever, presented in Fig. 4(b); (ii) the excellent DMT prediction of the Tabor parameter, $\mu < 0.1$, calculated after experimental determination of the radius of curvature R of the tip presented in Fig. 5, and of the pull-off force L_C presented in Fig. 6; (iii) the excellent DMT fit of the friction versus load experiments and the independent confirmation of the DMT fit by the experimental value of L_C presented in Fig. 6. Following the procedure suggested and supported above we found that a linear fit is the optimum fit for our friction force versus contact area representation in Fig. 7, demonstrating that, indeed, $F_f = \tau A$. Consistently, this free linear fit intercepts the origin, and the slope is a measure of the shear strength. We find that the shear strength $\tau = 238$ MPa, a value which lies within the typical range for AFM experiments (Carpick & Salmeron, 1997).

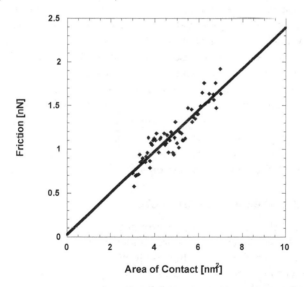

Fig. 7. Friction force versus contact area, showing a clear linear relation. The corresponding shear strength $\tau = 238$ MPa.

So, in contrast to the macroscopic law of friction, the friction force at the interface of a single-asperity is directly proportional to the contact area. Furthermore, since friction does not depend linearly upon the applied load for a single-asperity contact, one should be careful defining a friction coefficient, *i.e.* the friction force divided by the normal force, in AFM experiments, as its value varies with load.

The constant shear strength that we observe indicates that the mechanism of energy dissipation for this system does not change in this pressure range. Thus, the increase in friction with load is attributable to the increase in contact area, *i.e.* more atoms in contact, as opposed to a change in the frictional dissipation per interfacial atom. This may not be so surprising given that the nominal stress is only increasing as roughly $L^{1/3}$ (from the continuum mechanics models). The most likely mechanism of energy dissipation is thermalization of phonons generated at the contact zone during sliding. New modes of energy dissipation, resulting from inelastic processes, may activate at higher stresses (Carpick & Salmeron, 1997). For example, evidence of tip-induced atomic-scale wear has been reported for alkali-halide materials (Carpick et al., 1998b). Pressure-activated modes of energy dissipation are reported in organic thin films due to progressive molecular deformation (Barrena et al., 1999). These examples represent stress-dependent increases in the number of energy dissipation channels and are therefore manifested in increases in the shear strength compared with purely elastic, wearless friction.

Finally, we comment on the relative magnitude of the observed shear strength. The theoretical prediction for the shear strength of a crystalline material in the absence of dislocations is roughly given by $G/30$ (Cottrell, 1988), where G is the shear modulus. We can define an "effective" interfacial shear modulus $G_{eff} = 2G_{WC}G_{diamond}/(G_{WC} + G_{diamond}) \approx 380$ GPa . This gives, for the diamond/tungsten-carbide contact, $\tau \approx G_{eff}/1600$. The shear strength of this system is thus far below the ideal material shear strength (Hurtado & Kim, 1998). Previous AFM results of Carpick *et al.* (Carpick et al., 1996a; 1998b) and Lantz *et al.* (Lantz et al., 1997a; 1997b) observed shear strengths near the ideal limit. An ideal shear strength in the range of $G/30$ suggests a "crystalline" or commensurate interface that is free of dislocations, where the commensurability may be brought about by atomic displacements induced by interfacial forces. Our measured shear strength indicates that there may be very little atomic commensurability for the diamond/tungsten-carbide interface, which is plausible considering the high stiffness of these materials. More importantly, the hydrogen passivation of the diamond surface strongly reduces the adhesive force, and also the friction force. In fact, removal of the hydrogen passivation would result in a value for the shear strength which is much larger than the ideal theoretical prediction of $G/30$ (van den Oetelaar & Flipse, 1997).

3. Wear and third bodies in nanocontacts

We have investigated the nanotribological properties of a tungsten carbide tip in contact with a clean Pt(111) single crystal surface under ultrahigh vacuum conditions using scanning probe techniques. Because of the conductive nature of the cantilever and tip, we could alternate between contact atomic force microscopy (AFM) and non-contact scanning tunneling microscopy (STM) using the same probe. Several types of interfaces were found depending on the chemical state of the surfaces. The first type is characterized by strong irreversible adhesion followed by material transfer between tip and sample. This resulted in substantial amounts of material being transferred from the tip to the sample upon contact. This material often covered areas far exceeding that of the contact region. Low adhesion and no material transfer characterize a second type of contacts, which is associated with the presence of passivating adsorbates in both (full passivation) or in one of the two contacting surfaces (half-passivation). Half-passivated contacts where the clean side is the Pt(111) sample gave rise to periodic stick-slip friction behavior with a period equal to the atomic lattice constant of the Pt(111) surface. Local electrical conductivity measurements show a clear correlation between electronic and friction properties, with Ohmic behavior on clean regions of the Pt surface and semiconductor-like behavior on areas covered with adsorbates.

Our results indicate that substantial material transfer may be an important and inevitable property of nanocontacts when one surface is highly reactive and the other surface is not thoroughly cleaned. Furthermore, this work establishes that stable STM imaging using a conductive cantilever is a reliable method for observing this effect, and for observing fine features on clean portions of a reactive surface. In addition, the correlation between adhesion, friction, and contact conductance allows one to discern the existence and certain properties of the transferred material, which demonstrates that multi-functional scanning probe techniques are desirable for third-body processes at the nanoscale and nanotribology studies of tip-sample material transfer.

3.1 Background

The sliding of materials in contact often involves the transfer of material from one surface to the other. This material, referred to sometimes as the third body, influences the transient behavior of the sliding contact and can completely dominate the steady-state sliding behavior of many interfaces, especially for low friction coatings (Singer, 1992; 1998). Studies of low-friction materials such as diamond-like carbon coatings, MoS_2 coatings, and Ti-implanted steels indicate that chemically-modified transfer films are formed during initial sliding, and these films determine the long-term frictional behavior of the interface.

At small length scales third-bodies can also have a large impact on the contact properties. For example, hard disks and micro-electromechanical systems (MEMS) are critically limited by friction and adhesion-related failures due to the large surface-to-volume ratios of these devices (McFadden & Gellman, 1997). For such devices, an understanding of nano-scale third body behavior is important. Modeling work supports this notion. Robbins and co-workers have performed molecular dynamics simulations that indicate that molecular intermediate species in asperity contacts have a dramatic effect on friction (He et al., 1999; He & Robbins, 2001). They argue that contacts between crystalline or amorphous materials should, in general, exhibit very low friction due to the lack of interfacial lattice commensurability. The simulations show that molecules trapped at the interface, e.g. hydrocarbons, cause static friction that is consistent with observed macroscopic friction behavior.

The role of third bodies and transferred species at small scales is clearly worthy of further experimental study, specifically through the use of scanning force microscopy techniques. Already, fundamental insights into many aspects of friction have been obtained through the use of scanning force microscopy (Carpick & Salmeron, 1997). These studies have addressed several important topics such as atomic-scale stick-slip behavior, friction in the wearless (low-load) regime, friction in the presence of molecular lubricant films, the role of interfacial contact area, and wear initiation. However, there have been few studies of third body effects and transferred species. One example is the work by Qian et al. (Qian et al., 2000) who showed that in atomic force microscope (AFM) experiments, friction measurements exhibit transient behavior, where several tens of scans were required before friction behavior become reproducible. They proposed that the phenomenon is due to transfer of material between the tip and sample, and observed that the mechanism of transfer depended on the relative humidity and applied load. Carpick *et al.* (Carpick et al., 1996b) observed that the frictional shear strength and interfacial adhesion energy of a Pt/muscovite mica interface in ultrahigh vacuum progressively decreased with each scan, but recovered if the tip was "cleaned" by blunting it to expose fresh Pt. The authors suggested that potassium adsorbates transferred from the mica surface to the tip could explain the strong, progressive reduction of adhesion and friction observed in the experiment. Using the surface forces apparatus (SFA) Drummond *et al.* (Drummond et al., 2001) performed experiments where WS_2 nanoparticles were suspended in a tetradecane fluid, and then compressed and sheared between the two mica sheets of the SFA. They found that the nanoparticles formed a transfer film of nanometer-scale thickness on the mica that reduced friction appreciably.

In this work, we discuss the results of a study of a Pt (111)/tungsten carbide single asperity interface using a combination of ultrahigh vacuum (UHV) AFM and scanning tunneling microscopy (STM) techniques. Since both the Pt sample and the carbide tip are conductive, we were able to measure the local electrical conductance of the contact and the friction force simultaneously (Enachescu et al., 1998; 1999a). In addition, the conducting tip allows STM operation, whereby high-resolution non-contact images of the sample can be obtained before and after the contact experiments.

3.2 Experimental

The experiments were performed in a UHV chamber (base pressure 7×10^{-11} Torr), equipped with a home-built AFM (Dai et al., 1995), low-energy electron diffraction (LEED), Auger electron spectroscopy (AES), differentially pumped ion sputtering, and sample cooling and annealing capabilities. The Pt(111) single-crystal sample was cleaned by sputtering with Ar+ ions of 1 keV energy, both in hot conditions (600 ^{0}C) and at room temperature, for 10-20 min. After sputtering O_2 was introduced in the chamber at 10^{-6} Torr for ~3 min while the sample temperature was kept at 600 ^{0}C. O_2 exposure and heating were then stopped. The cycle was repeated two or three times. Finally, the sample was flashed to 950-1000 ^{0}C for ~1 min. and cooled down at ~2 ^{0}C/sec. The AES pattern taken after the cleaning procedure indicated a clean Pt(111) surface, with the carbon peak at 271 eV not visible above the noise level, while platinum peaks at 237 eV and 251 eV are clear visible (Fig. 8). A sharp (1x1) LEED pattern was also observed (inset, Fig. 8).

Commercially available triangular silicon cantilevers with integrated tips, coated with approximately 20 nm of tungsten carbide, were used for all measurements. The cantilevers were characterized by scanning electron microscopy (SEM) in order to determine tip and lever dimensions, and also by AES to determine the chemical composition of the lever and tip shaft. The measurements showed the presence of both tungsten oxide and carbide, which is not uncommon for such coatings. For convenience we will refer to these as "tungsten carbide" tips. The similar chemical composition of the lever and tip is quite normal, as the WC coating is covering not only the tip but also the cantilever. For *such conditions* one may suggest that tip cleanliness is similar to the cantilever cleanliness, i.e., both covered with tungsten oxide and carbide. However, for our AFM/STM measurements most of the time the tip cleanliness is *not* similar to the cantilever cleanliness, as we often clean the AFM-STM tip and thus, removing the tip contaminant.

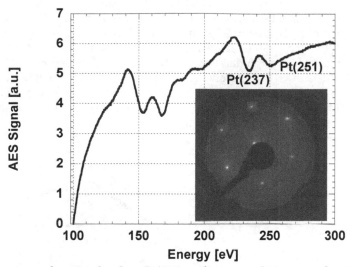

Fig. 8. AES spectrum showing the clean Pt(111) surface, e.g., platinum peaks at 237 eV and 251 eV are clear visible . Inset: a LEED pattern displaying the a clear Pt(111)-(1x1) pattern.

Two cantilevers with different spring constants of 88 N/m and 3.1 N/m were used. The stiffer cantilever was used for conductance and friction measurements while the other was used for certain high resolution friction measurements only. The tips were cleaned in UHV immediately prior to the measurements, by applying short voltage pulses and/or by rubbing them against the surface. Normal cantilever force constants were taken from the manufacturer, and the normal/lateral force ratio was calculated using the method described in (Ogletree et al., 1996). The absolute accuracy of the forces measured is limited due to significant uncertainty in the material properties of the cantilever and approximations used in the force constant calculations. However, relative changes in friction could be accurately determined by using the same cantilever and tip during a series of measurements. A large dynamic range, two-stage I-V converter was built, which provided a large frequency range while achieving sufficient gain. For this work, lower I-V gains (*e.g.*, 10⁴) were used to measure the current flowing through the tip-sample junction while in AFM-contact mode, higher gains (*e.g.*, 10⁹), were used for tunneling microscopy using the AFM tip.

3.3 Results and discussions

3.3.1 Irreversible adhesion between clean interfaces

As mentioned above, the Pt(111) sample was cleaned by sputtering and annealing, and its state checked by AES and LEED to verify the chemistry and structure of the surface. Because of the nanoscale dimensions of the tip apex, we could not assess spectroscopically its chemical state in the UHV chamber. However, we found that scanning at high loads on sacrificial areas of the sample consistently produced tips with highly adhesive properties and metallic conductance characteristics. Thus, rubbing a contaminated WC tip on a *clean metal* surface is an effective way to clean the tip. The adhesion force for tips prepared in this manner was large enough that even at the lowest load scanning was not possible without severe damage. These contacts were characterized by means of force-displacement curves, as in the example shown in Fig. 9, where a pull-off force of L_c = 12.0 ± 1.2 µN was measured with the cantilever of 88 N/m normal spring constant. Assuming, for simplicity, an *elastic* adhesive contact, this force can be related to the work of adhesion of the interface. Within the extremes of the Johnson-Kendall-Roberts (JKR) model (Johnson et al., 1971), and the Derjaguin-Muller-Toporov (DMT) model (Derjaguin et al., 1975), we obtain an "effective" work of adhesion between 12 and 16 J/m². For this calculation, we used a value of 160 ± 20 nm for the tip radius, which was measured experimentally by scanning over sharp edges of a faceted $SrTiO_3$(305) sample (Carpick et al., 1996a).

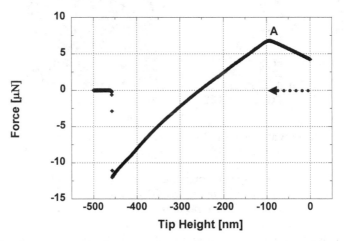

Fig. 9. Force-displacement curve for a tungsten carbide tip in contact with a clean Pt(111) sample. The plot is shown for the retracting portion only. The load appears to increase when retraction starts, even thought the tip-sample separation is increasing. This is due to the strong adhesion of the tip to the surface, which prevents sliding. Consequently, the tip pivots about the contact point. After that, the load decreases down to the pull-off point, as is usual in force-displacement curves. The deformations of the cantilever giving rise to this behavior are illustrated in Fig. 10.

This adhesion energy is likely an overestimate because we have neglected the possibility of plastic failure of the junction between the tip and sample. It is extremely difficult to apply such a model to this contact without knowing more about the contact geometry or the species at the interface. In any event, this effective work of adhesion is three orders of magnitude higher than that found in previous UHV AFM measurements (Carpick & Salmeron, 1997), such as 0.02 J/m² between silicon nitride AFM tips and the muscovite mica surface in UHV, or 0.4 J/m² for a Pt tip on a mica surface in UHV (Carpick et al., 1996b). Note that the surface energy of most metals is in the range of 1 to 5 J/m², i.e., 2 to 10 J/m² are required to split an ideal crystal in half to create two new surfaces (Israelachvili, 1992). Our value of 16 J/m² is beyond this range, consistent with the notion that we have likely overestimated the adhesion by assuming elastic contact. Nevertheless, our measurement indicates that we have observed extremely strong adhesion for this pair of materials. This indicates that strong bonds, at least several eV per atom suggestive of covalent bonds, are formed between the clean Pt(111) surface and the tungsten carbide tip.

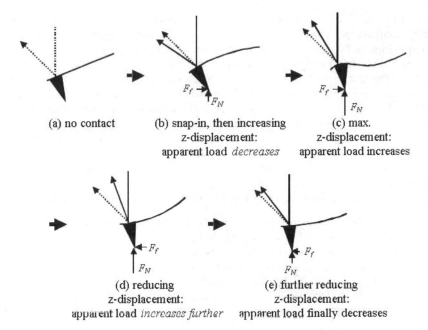

(a) no contact

(b) snap-in, then increasing
z-displacement:
apparent load *decreases*

(c) max.
z-displacement:
apparent load increases

(d) reducing
z-displacement:
apparent load *increases further*

(e) further reducing
z-displacement:
apparent load finally decreases

Fig. 10. Schematic drawing illustrating the deformation of a cantilever subject to high friction forces. (a) The cantilever is shown at the upper left for the case of no load. The dashed line represents the trajectory of the laser for the zero load case, and is included in all subsequent sketches. (b) After snapping into contact, the z-displacement is increased. Normally this would cause an increase in the laser signal (after the initial decrease due to the snap-in). But large friction forces cause bending at the end of the lever in the opposing direction, and producing an apparent *decrease* in the load. Friction is preventing the tip from sliding relative to the surface. (c) As the z-displacement is further increased, the friction force eventually reaches its limiting value. The tip will begin to slide relative to the surface, and the upward bending induced by the z-displacement overtakes the tendency to bend in the opposite direction induced by the friction force, so now the apparent load begins to increase. Eventually the z-displacement reaches a maximum value. (d) As the z-displacement direction is reversed, the friction force will now resist motion in the opposite direction, and so the bending it induces causes an apparent increase in the load. During this phase, the tip is not sliding relative to the surface. (e) Eventually the friction force reaches a limiting value and once again the tip begins to slide. The apparent load will now decrease as the z-displacement is decreased. Stages (d) and (e) are clearly evident in Fig. 9.

In addition, an unusual hysteresis feature in the force-displacement plots was observed at the largest z-displacements as seen in Fig. 9. The recorded data begins with the tip initially in contact after being pushed back by approximately 450 nm. As the cantilever is retracted, the apparent force on the cantilever *increases*, then eventually begins to decrease, as one would normally expect. This result can be explained by considering the effect of friction on the cantilever bending due to the tilted geometry with respect to the plane of the sample (22.5° in this case). Friction causes the cantilever to bend in addition to bending due to the normal force between the tip and sample. The direction of the bending will depend on the direction of the friction force, which always acts to oppose any tendency for sliding. These two forces compete by changing the slope of the cantilever in opposite directions. This is illustrated in Fig. 10. Because of the strong bonding, the tip cannot slide over the surface and the cantilever is forced to adopt an S-shape like the one shown in Fig. 10(c). As the sample is retracted and the lever reverts to its normal bending shape (shown in Fig. 10(d)), it produces an apparent increase in the force initially. After passing through a maximum (point A in Fig. 9), the force decreases as expected. This effect is explained by the strong friction force on a cantilever fixed at one end and with a tilted geometry with respect to the plane of the sample (22.5° in this case). The slope of the force-displacement curve is inverted because static friction prevents the tip from sliding with respect to the surface. Instead the tip is pivoting about the contact point, and the slope of the end of the cantilever is increasing. Eventually the tip pivots enough that the maximum static friction force is reached and the tip can slide relative to the surface. Stages (d) and (e) sketched in Fig. 10 are evident in Fig. 9.

The fact that this hysteresis phenomenon is observed for a very stiff lever, *i.e.* a spring constant with two to three orders of magnitude higher than the typical contact AFM levers (88 N/m), indicates that strong friction forces are occurring in tandem with the strong adhesion forces. In addition, our experiments show a much stronger interaction between the AFM tip and the metallic surface compared to the results published by Bennewitz *et. al.* (Bennewitz et al., 1999; 2001). Those measurements involved silicon AFM levers having a spring constant of 0.024 N/m in contact with clean a Cu(111) surface under UHV conditions.

3.3.2 Contacts between fully passivated interfaces: Friction and conductance measurements

As we have seen with clean tips and clean surfaces, contact-mode measurements cannot be performed without severely disrupting the contact region. To perform contact experiments while avoiding strong modifications, the surfaces must be chemically passivated. This can be achieved intentionally or unintentionally by the presence of adsorbate layers. In our case we used the unintentionally passivation, provided without effort by the contaminant's presence, i.e., the contamination behaved like a "passivation" layer for the tip-sample interaction. An interesting question is whether these layers must be present on each or on only one of the surfaces for substantial passivation. The latter case implies that the layers are attached strongly to one of the surfaces and interact only weakly with the other, such that the contact can shear at this weak interface. As we will show in this and the next section, it is indeed possible to have both situations, which we shall call fully passivation when layers of material are present on each contacting surface, and half passivation, when one of the two surfaces remains clean, during and after friction scanning.

On the Pt surface, the most common contaminant after annealing in UHV is carbon, as verified with AES. On the WC tip, in addition to oxygen present as a tungsten oxide, adventitious hydrocarbon or graphitic carbon can also accumulate. Ex-situ AES on the body of the cantilever and on the tip shaft did indeed reveal the presence of O, C and W as the only observable constituents. We will first examine results where the Pt surface is covered by a layer of C-contamination, the only impurity element detected in the Auger spectra. Force-displacement data obtained with such passivated surfaces show low adhesion values in the range of ~1 J/m^2, depending on the spatial location of the tip over the surface, as shown in the example of Fig. 11.

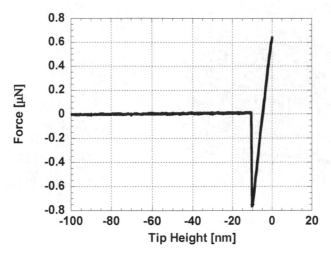

Fig. 11. A force-displacement plot for the same lever as in Fig. 7, but taken over a passivated area of the Pt(111) sample. The adhesion force is much lower, and the unusual behavior in the plot due to friction forces is not present.

Topography, friction, and point contact current (corresponding to contact conductance) were recorded simultaneously as the tip was scanned over the surface, (Enachescu et al., 1999b) as shown in Fig. 12(a,b). The friction and conductivity maps exhibit a strong correlation, with regions of high and low friction corresponding to regions of high and low electrical conductivity respectively.   We propose that high friction and high local conductivity are associated with "cleaner" interfaces, while low friction and low local conductivity correspond to regions covered with more interfacial adsorbates. The spatial distribution of friction and conductance values remained consistent throughout several images. This indicates that the tip was not changing during the image acquisition, but rather, different regions of the sample had different amounts of adsorbates present.

Typical I-V characteristics obtained with contacts in areas with different degree of passivation are shown in the graphs of Fig. 12(c). We acquired current-voltage (I-V) curves in a 16 x 16 grid, while for each I-V curve the voltage was scanned from -50 mV to +50 mV. Ohmic behavior curves (straight lines) were always observed in the regions exhibiting high friction, while semiconductor-like behavior (sigmoid shapes) were observed in low friction areas. This observation can be understood on the basis of the poor conductivity of the contamination layers present, which decreases as their quantity, and thus passivation capacity, increases.

3.3.3 Contacts with half-passivated interfaces: Atomic lattice resolution images

In contrast with the fully passivated interfaces, when the Pt surface is clean, we could frequently observe stick-slip behavior with the atomic-lattice periodicity of the Pt(111) substrate, as shown in the image of Fig. 13(a) and the trace of a friction line in Fig. 13(c). The Fourier transform of the image (Fig. 13(b)) shows more clearly the 3-fold symmetry with 0.27 nm periodicity, in agreement with the lattice constant of Pt(111). The image was acquired under zero externally applied load. The occurrence of stick-slip behavior was always

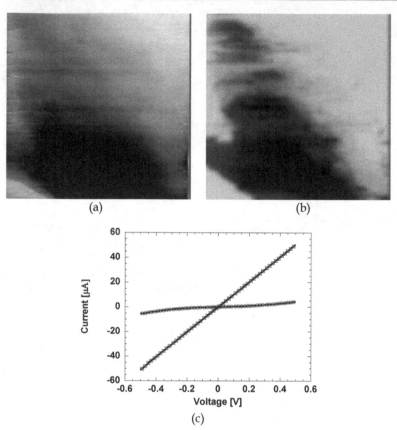

(a)   (b)

(c)

Fig. 12. Simultaneous friction (a) and point contact current (b) images of a Pt(111) surface acquired with a conductive WC-coated Si cantilever of 88 N/m spring constant. Image size is 500 x 500 nm². Regions with high and low friction are clearly correlated to regions of high and low local conductivity. (c) Corresponding *I-V* spectra acquired at the points of "clean" and "passivated" areas. The bias between tip and sample was varied from -50 mV to +50 mV. High friction regions are correlated with Ohmic conductance behavior, while the lower friction regions exhibit non-Ohmic conductance, indicative of an insulating or semiconducting interlayer.

associated with the presence of low adhesion, low friction, and low contact current, indicative of chemically inactive tip. These atomic stick-slip images were recorded by carefully choosing flat terrace locations. We did not scan areas that included steps, in order to avoid the increased reactivity of the step sites.

This result indicates that the passivating layers are on the tip side of the interface, where they are bound strongly enough to withstand the applied shear stresses without transference to the Pt surface. In contrast to these unintentional adsorbates used as passivation layer, other intentional, well-defined adsorbates may not be well bound as to the WC tip and thus, may not provide any additional understanding to the tip-metal interaction we described in this work. An analysis of the energy balance during friction is of interest here.

Since the friction force is approximately 190 nN, the energy dissipated after a displacement of one unit cell (~3 Å) is about 350 eV. Given the tip radius of 160 nm, and a total load of ~0.8 μN (see Fig. 11) a contact area of roughly 100 nm² can be calculated containing approximately 1000 atoms. This corresponds to an average energy dissipated per atom of 0.3 eV. We would predict that this energy is not enough to break the strong chemical bonds of the Pt atoms, and indeed that is what we observe.

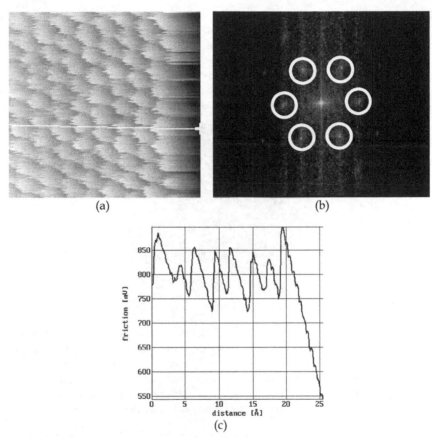

Fig. 13. (a) 2.5 x 2.5 nm² lateral force image obtained with a passivated tip and a clean Pt(111) sample. Atomic-lattice stick-slip friction is clearly observed and provides proof that sliding is taking place along a clean Pt surface; (b) Fast Fourier transform of the data in (a), showing the periodicity and symmetry of the Pt(111) surface; (c) line trace of the line indicated in (a) showing the clear stick-slip behavior.

3.3.4 Tunneling experiments using the AFM tip

In most cases the adhesion between the AFM tip and the Pt(111) was so strong that we were not able to scan the tip over the sample and often the fracture of low spring-constant AFM cantilevers was observed. Reproducible scanning over the Pt(111) surface was only possible when a small amount of contaminant was present between the tip and the sample.

To further investigate the Pt(111) surface we performed non-contact scanning tunneling microscopy (STM) experiments using the conductive WC AFM tip. For these experiments the AFM cantilever must be sufficiently stiff, otherwise the cantilever will jump into contact (Carpick & Salmeron, 1997; Enachescu et al., 1998) before or during tunneling conditions at small physical gaps, *e.g.*, 0.7-1.0 nm. We used an 88 N/m stiff lever, which was enough to avoid the jump to contact.

Fig. 14 shows an STM image acquired with an AFM lever prior to any tip-sample contact. The 200 nm x 200 nm image was acquired with a bias of 0.1 V (sample negative) and a tunneling current of 160 pA. Several monoatomic steps, with a height of 0.22 ± 0.01 nm, and terraces are visible. A few isolated protrusions are also observed on the terraces and also attached to the step edges. They correspond probably to contamination. The round flat islands attached to the steps have a height of 0.1 nm and clearly pin the steps, which would have been flowing during the high temperature anneal.

Fig. 14. 200 x 200 nm² STM image of the Pt(111) surface taken *before* any contact has been made between the tip and sample. A series of single atom steps, 0.22 ± 0.01 nm in height, are observed. Two of the steps are seen to be pinned by contamination, which is likely carbon that has diffused to the surface region from the bulk during processing of the crystal. The image was acquired for an I-V converter gain of 10^8 under the following tunneling condition: tunneling bias of 0.1 V (sample negative) and a tunneling current of 160 pA.

Fig. 15 shows a 200 x 200 nm² STM image acquired with the AFM lever *after* mechanical contact has been made between the tip and sample. Specifically, the tip was brought into contact with the sample and then retracted. The tip was not scanned during this contact. The tip was then brought back to within tunneling range and used to acquire the STM image with a bias of -0.2 V and a tunneling current of 160 pA. Large, irregular features up to 13.7 nm in height are seen covering roughly 75% of the image. In the topmost 25% of the image Pt(111) steps can still be seen. This image demonstrates that the bonding of the tip to the surface was very strong, and that rupture of the contact occurred within the tip material itself. This material is then left over the Pt surface.

Another example of STM imaging followed by contact and AFM imaging and then again STM imaging is shown in Fig. 16. The 100 nm x 100 nm image in (a) was acquired at a bias of -0.2 V and tunneling current of 200 pA and shows a stepped region of the Pt(111) surface

Fig. 15. STM image of the Pt(111) surface after AFM mechanical contact has been made between the tip and sample. The 200 x 200 nm² STM image was acquired with a tunneling bias of -0.2 V and a tunneling current of 160 pA, while the I-V converter was operating at a 10⁸ gain.

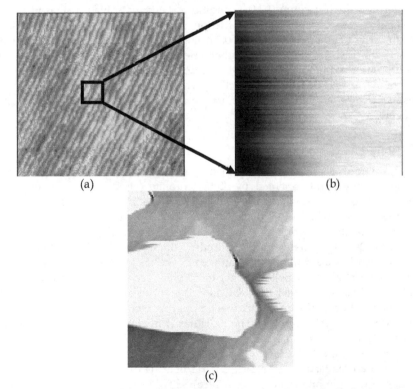

(a)   (b)

(c)

Fig. 16. (a) 100 x 100 nm² STM image of a highly stepped region of the Pt(111) surface. The image was acquired under the following tunneling conditions: bias = -0.2 V; current = 200 pA; (b) 10x10 nm² AFM friction force image acquired in contact at the center of (a); (c) Subsequent STM image, acquired under the same conditions as in (a), showing substantial material deposition from the AFM tip during the previous contact before contact. Fig. 16(b) shows a 10x10 nm^2 AFM contact friction force image at zero externally applied load acquired in the center of Fig. 16(a). During imaging the bias was held at zero volts. High friction forces are observed in this image. The tip was then withdrawn from contact and another STM image acquired, shown in Fig. 16(c), under identical tunneling conditions as Fig. 16(a). It is obvious from Fig. 16(c) that substantial material deposition from the AFM tip during contact scanning process has taken place.

Using the friction force during slip of several µN, the energy dissipated during imaging can be estimated to be of the order of 10^8 eV. The amount of material left on the surface (Fig. 16(c)) occupies an area of roughly 2,500 nm^2. The energy dissipated during friction is therefore sufficient to break the junction, even if very strong (~5 eV) bonds need to be broken.

4. Local conductivity in nanocontacts: The integration of point-contact microscopy

The electrical current through the point-contact junction of an AFM tip was used to image the surfaces of bulk graphite (HOPG) and the surface of a graphitized carbon monolayer on Pt(111) under ultra-high vacuum (UHV) conditions. Lattice-resolved images were obtained simultaneously in topography, lateral friction, and contact current channels. Lattice resolution in current maps persisted up to 0.9 mA and pressures of up to 5 GPa. In both bulk graphite and the case of graphitized carbon monolayer on Pt(111), the current images showed only one maximum per unit cell. In addition, the contact current images of the graphite monolayer revealed local conductivity variations. We observed local conductivity variations in the form of moiré superstructures resulting from high order commensurability with the Pt lattice.

4.1 Background

Since the invention of the scanning tunneling microscope (STM) (Binnig et al., 1983), graphite, specifically highly-oriented pyrolytic graphite (HOPG), has become a popular substrate due to its flat cleavage surface and its inert nature, which makes it possible to obtain images in air with "atomic resolution" (Hansma, 1985) (Binnig et al., 1986b). However, the literature reports a number of well-known puzzling features, such as uncharacteristically large corrugation amplitudes (Binnig et al., 1986; Selloni et al., 1985; 1986; Tersoff, 1986; Batra & Ciraci, 1988; Batra et al., 1987; Soler et al., 1986), enhanced lateral resolution (Binnig et al., 1986; Selloni et al., 1985; 1986; Tersoff, 1986; Batra & Ciraci, 1988; Batra et al., 1987; Soler et al., 1986; Park & Quate, 1986), a weak dependence of the tunneling current on the position of the tip in the direction perpendicular to the surface (Salmeron et al., 1991), and anomalously large superperiodicities (Kuwabara et al., 1990). These features generated a debate about the imaging mechanism. In most STM images, one observes only one maximum per unit cell, indicating that the carbon atoms are not imaged as individual units. In a favored explanation, the lattice periodicity is due to the tip imaging a single electron state of the graphite layer (Tersoff, 1986). STM images taken on one monolayer of graphite deposited on metals also show only the lattice periodicity, and not single atomic positions (Land et al., 1992a; 1992b).

In the debate concerning the imaging mechanism of HOPG in STM, it was suggested that the STM tip could be in contact with the HOPG. In order to clarify this issue, Smith *et al.*

(Smith et al., 1986) performed an experiment imaging the HOPG surface by purposely placing the tip in contact with the surface. In contact, the situation is similar to that in point-contact spectroscopy (Yanson et al., 1981). This mode of microscopy was called point-contact microscopy (PCM) (Smith et al., 1986), which differs from STM in that the tip is much closer to the sample in the region where the potential barrier is significantly reduced and tip-sample forces are repulsive. In this mode, Smith *et al.* succeeded in imaging the HOPG lattice by measuring the current flowing through the contact. However, they were able to report lattice resolution only at low temperatures, *i.e.*, when the microscope was immersed in liquid helium. Since their contact area involved a large number of atoms, they explained the "atomic" resolution by considering the conduction to be due to a single atom on the tip, which we now consider unlikely. Other experimental evidence indicates that true tunneling through a vacuum gap might not occur in the case of graphite in normal circumstances, and that the tip is in contact with the surface (Salmeron et al., 1991).

In AFM contact mode, lattice resolution can be obtained both in topography and friction channels. This is usually explained as the result of stick-slip phenomena (Marti et al., 1987). AFM measurements involving a conductive lever have been reported (Lantz et al., 1997a, 1997b; Enachescu et al., 1998) in conjunction with tip-sample contact area evaluation. To date, there have been no reports of AFM contact experiments on graphite deposited on metals.

This work presents results on the simultaneous implementation of AFM and PCM techniques by using a conductive AFM lever. We demonstrate the possibility of obtaining lattice resolution concurrently in three channels: topography, friction, and contact current. This is achieved by using both HOPG and 1 ML of graphite deposited on a Pt(111) single-crystal. We show that PCM is as capable of similar lateral resolution as contact AFM imaging. We also found that PCM is sensitive to local conductivity variations due to moiré superstructures resulting from the high order commensurability of the graphite and Pt lattices at different relative rotations. Moreover, we show that lattice resolution in PCM mode is achievable for currents of up to 0.9 mA and contact pressures estimated at 5 GPa.

4.2 Experimental

All experiments were performed in a UHV chamber (base pressure 7×10^{-11} Torr) equipped with AFM, Auger Electron Spectroscopy (AES) and Low-energy Electron Diffraction (LEED) (Dai et al., 1995). Two different samples, HOPG and Pt(111), were used. The HOPG sample was cleaved along the (0001) plane in air and then immediately placed in the vacuum chamber.

The samples could be heated by means of electron bombardment from a hot dispenser cathode. The Pt sample was prepared using standard procedures of Argon ion bombardment, oxygen treatment, and annealing until a clean and ordered surface was produced, as verified by AES and LEED. The clean surface was then exposed to ethylene at room temperature by backfilling the chamber with ethylene. Exposures were typically greater than 10 Langmuir to ensure saturation of the Pt(111) surface. After exposure, the sample was heated to about 1250K, resulting in the decomposition of ethylene and formation of a single monolayer of graphite on the Pt(111) surface. When observed with LEED, we found that the graphite layer produced characteristic fragmented rings (Hu et al.,

1987), with several dominant bright segments. Some of the ring segments were in-line with the Pt spots, indicating alignment or near-alignment of the graphite and Pt lattices. Others were at an angle relative to the Pt spots, indicating that the graphite lattice was rotated with respect to the Pt lattice.

We measured derivative Auger spectra of the surface with an RFA-type electron analyzer, using a normally incident electron beam with an energy of 2500 eV and retarding field oscillation amplitude of 7 eV peak-to-peak. The ratio of the peak-to-peak heights of the carbon (275 eV) and platinum (237 eV) AES transitions in the derivative spectrum was found to be about 3.8, independent of the amount and method of ethylene exposure. We attribute this to saturation of the surface once a graphite monolayer is formed, such that no further decomposition of ethylene can take place. Because of this, we concluded that there was 1 ML of graphite on the Pt(111) surface. This conclusion is supported by Land et al. (Land et al., 1992a, 1992b), who determined by STM, under similar preparation conditions, that the deposited graphite layer was 1 ML thick.

All experiments were performed with a silicon cantilever with a spring constant of 3.5 N/m, and coated with a ~20 nm thick conductive tungsten-carbide layer (bulk resistivity ~30 $\mu\Omega$-cm). The tips were characterized by Scanning Electron Microscopy (SEM) and AES. Previous UHV-AFM measurements on a Pt(111) sample showed that similar tungsten-carbide coated tips are wear resistant and conductive (Enachescu et al., 1998, 1999b). The tips were treated in UHV immediately prior to the measurements by applying short voltage pulses while in contact and/or by rubbing them at high loads on the surface. We designed and built a flexible I-V converter that allowed us to measure high contact currents by taking measurements spanning the range from pA to mA.

4.3 Results and discussions

4.3.1 Topography, friction, and current imaging

Fig. 17 shows three 2.5 nm x 2.5 nm images of the HOPG surface, which were acquired simultaneously. The feedback control was turned off in order to avoid the convolution of topography and friction, and to minimize noise. The images correspond to: (a) normal lever deflection, (b) lateral force or friction, and (c) contact current. A positive bias of 1.0 V was applied to the sample, and the external load during imaging was 100 nN. The average current in (c) was 0.94 μA, with a modulation of about 17%. In all three images, the 0.246 nm graphite lattice periodicity is clearly observed. Using the DMT contact mechanics model (Müller et al., 1983; Derjaguin et al., 1975), with a measured pull-off force of 115 nN, we estimate that our contact radius is 4.1 nm, and therefore contains about 2000 atoms. The contact radius calculated here is only approximate, since the Tabor parameter (Greenwood & Johnson, 1998; Tabor, 1977) for this system is 0.67, which indicates that the DMT model is not entirely appropriate. Moreover, none of the analytical contact mechanics models are directly applicable to a non-isotropic material such as graphite (Sridhar et al., 1997).

Similar lattice-resolved images were obtained on 1 ML of graphite on Pt(111) (Gr/Pt(111)). An example of this is shown in Fig. 18. As in the previous case, the 2.5 nm x 2.5 nm images correspond to: (a) normal lever deflection (under feedback-off conditions), (b) lateral force, and (c) contact current. The external load in this case was 300 nN, and the sample bias was 0.5 V. The average current was 52.7 μA, with a current modulation of about 2%. In this case,

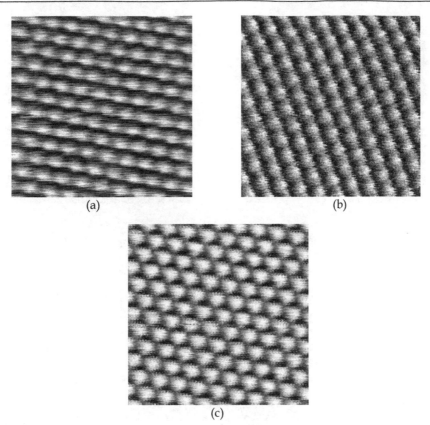

(a)
(b)
(c)

Fig. 17. Simultaneously acquired lattice resolution images of HOPG under UHV conditions: (a) normal lever deflection (with topographical and buckling effects), with a corrugation of 117 pN, corresponding to a height of 33.6 pm; (b) lateral friction image, average force of 0.5 nN and corrugation of 20 nN; (c) PCM image, average current of 945 nA and peak-to-peak corrugation of 160 nA. Image was taken with an applied load of 100 nN without feedback. Image size is 2.5 nm x 2.5 nm.

the diameter of the area of contact was similarly estimated to be 5.78 nm, which contains approximately 4000 atoms. Here the 0.246 nm graphite lattice periodicity is also clearly revealed. We were able to obtain lattice resolution at currents up to 0.9 mA and high load. The average pressure at high load was approximately 5 GPa, which is less than the theoretical yield stress of Pt (~17 GPa). At pressures higher than 5 GPa and/or currents higher than 0.9 mA, the images were unstable, although the graphite lattice was still visible.

As a side note, we found that we were able to obtain lattice resolution almost all the time and immediately in current mode, while lattice resolution was not as readily visible in the topography and/or friction channels. In many cases, the friction was so low that there was no stick-slip present, i.e., the tip moved continuously over the graphite layer. Because of these reasons, we can rule out the atomic stick-slip mechanism as a reason for the lattice resolution observed in PCM mode.

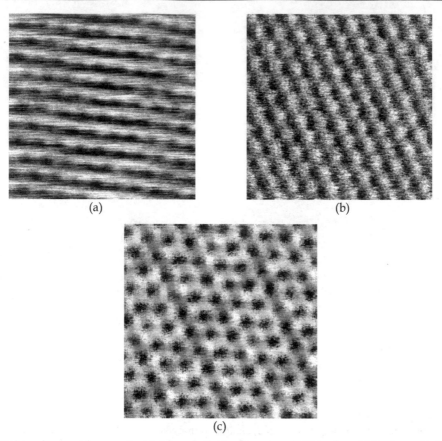

(a)   (b)

(c)

Fig. 18. Simultaneously acquired lattice resolution images of 1 ML of graphite deposited on Pt(111) in UHV: (a) normal lever, corrugation of 164 pN; (b) lateral friction image, average force of 0.4 nN and corrugation of 17 nN; (c) PCM image, average current of 53 μA and peak-to-peak corrugation of 1.1 μA at a bias of 0.53 V. Image was taken with an applied load of 300 nN without feedback. Image size is 2.5 nm x 2.5 nm.

4.3.2 Moire´ structures

It is known that, for similar preparation conditions (Land et al., 1992a; Hu et al., 1987), graphite forms several orientational domains on a Pt(111) sample. Depending on the preparation conditions and annealing temperature, different sizes and orientations of domains can be prepared. In Fig. 19, we show a 60 nm x 60 nm image of two graphite domains on Pt(111).

The hexagonal periodicity observed in the upper left domain in this image is about 2.0 nm. The large unit cell arises from the superposition of the incommensurate lattices of graphite and Pt(111) at a particular angle. In higher resolution images of this domain, such as the one shown in Fig. 20, the graphite lattice of 0.246 nm, together with the larger 2.0 nm cell, is revealed. Using the real space image and its 2-D Fourier transform, we find that the

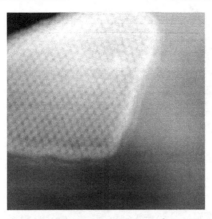

Fig. 19. PCM image showing two moiré superstructure domain on graphite/Pt(111). The upper left has a periodicity of ~2.0 nm, and the superstructure lattice was determined to be ($\sqrt{63}$ x $\sqrt{63}$)R19 with respect to the graphite lattice. Image size is 60 nm x 60 nm. The other domain in this image is (5 x 5) with respect to the graphite lattice, although it is not resolved at this scale.

Fig. 20. Close-up image of the ($\sqrt{63}$ x $\sqrt{63}$)R19 domain in Fig. 19, showing the graphite lattice, as well as the moiré superstructure. Image size is 10 nm x 10 nm. Average current is 90 μA and corrugation is 5 μA at a bias of 0.8 V.

structure in Fig. 20 corresponds to a superstructure with a ($\sqrt{63}$ x $\sqrt{63}$)R19 unit cell with respect to the graphite lattice. Contrary to standard usage, we report on the moiré structures with respect to the overlayer instead of the substrate, since we can directly count the number of graphite unit cells in the moiré superstructure. Using the known lattice constants of graphite and Pt and the measured angles, we can calculate that the moiré periodicity is almost exactly 7 Pt lattice spacings, and the moiré cell is rotated by 22° with respect to the Pt lattice. Indeed, one can create the 2.0 nm periodic superstructure by rotating the Pt [1-10] direction with the graphite [1010] direction by 2.68°, as shown in Fig. 21. There is a small lattice misfit of 0.60 % associated with the coincidence of the graphite

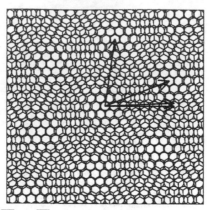

Fig. 21. Schematic of the ($\sqrt{63}$ x $\sqrt{63}$)R19 (with respect to graphite) moiré domain superstructure. With respect to the Pt(111) substrate, the moiré domain is (7 x 7)R22. The Pt atoms are shown as circles with a scaled diameter equal to the lattice constant of Pt (0.278 nm). The graphite lattice is shown as hexagons in which carbon atoms are located at the vertices with C-C distance of 0.142 nm and lattice constant of 0.246 nm. Vectors are drawn to indicate the orientation of the two lattices and the moiré domain. Image size is 5.5 nm x 5.5 nm.

lattice at 1.954 nm and the Pt lattice at 1.942 nm, which can be accounted for by a corresponding relaxation of the graphite layer or the platinum substrate.

Other graphite domains having different orientations and moiré superstructures have been observed, frequently adjacent to each other. The image in Fig. 22 shows three contiguous graphite domains, each having different orientations. It is interesting to note that the average current in these domains is different, even if all other conditions (bias, load, tip structure) are the same. The average current can also vary appreciably inside a single domain, such as at a platinum step, as we discuss below.

Fig. 22. PCM image showing different moiré domains. Image size is 100 nm x 100 nm. Note that the average current is different on each domain. Average currents are 86, 100 and 97 μA for the left, center and right domains, at a bias of 0.8 V.

At higher magnification, different periodic superstructures on each domain can be seen. The image in Fig. 23, which was obtained from the left domain of Fig. 22, shows a (3 x 3) modulation of the graphite lattice. Its 2-D Fourier transform is shown in Fig. 24. The larger hexagonal pattern, marked by six circles, corresponds to the 0.246 nm graphite lattice, while the smaller hexagon, marked by squares, represents the 0.738 nm superstructure lattice. The calculated angle of the graphite lattice with respect to Pt(111) lattice is 19.1°, which is in agreement with the measured angle.

Fig. 23. PCM image of a (3 x 3) moiré superstructure, showing the graphite lattice. Image size is 5 nm x 5 nm. Average current is 79 µA and modulation amplitude is 0.93 µA at a bias of 0.7 V.

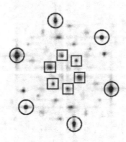

Fig. 24. Fourier transform of image in Fig. 23, showing the graphite lattice periodicity of 0.246 nm marked by circles, and the moiré superstructure periodicity of 0.738 nm marked by squares.

The domain in the middle of Fig. 22 has a (5 x 5) modulation of the graphite periodicity, as shown in the 5 nm x 5 nm image of Fig. 25. In this moiré structure, the angle between the graphite and the Pt(111) lattices is calculated to be 23.4°. The domain in the lower right of Fig. 22 was identified as a ($\sqrt{31}$ x $\sqrt{31}$)R9 structure.

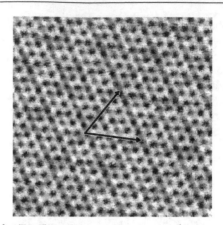

Fig. 25. Close-up image of a (5 x 5) moiré superstructure, showing the graphite lattice. The arrows indicate the moiré lattice. Image size is 5 nm x 5 nm.

Table 1 is a list of the experimentally observed moiré structures. Using the ratio of the lattice constants of graphite and Pt, we were able to calculate near-coincidences of the graphite and Pt lattices at different angles, and thus predict the existence of all of the structures.

Moiré periodicity relative to graphite	Moiré periodicity relative to Pt(111)	Angle between graphite and Pt lattices [^{0}]	Moiré periodicity [nm]	Coincidence misfit [%]
(3x3)	(√7x√7) R19	19.1	0.738	0.60
(√19x√19) R23	(4x4)	23.4	1.07	3.4
(5x5)	(√19x√19) R23	23.4	1.23	1.7
(√31x√31) R9	(5x5)	8.9	1.37	1.2
(√52x√52) R14	(√43x√43) R8	21.5, 6.3	1.77	2.5
(√61x√61) R26	Unknown		1.92	< 4.0
(√63x√63) R19	(7x7) R22	2.7	1.95	0.60
(√73x√73) R6	Unknown		2.10	< 4.9

Table 1. Moiré superstructures experimentally observed in PCM mode. Structures with respect to the Pt lattice are deduced from the measured structures on graphite. In certain cases, the angle of the Pt lattice was known, which aided in the extrapolation.

4.3.3 Measuring local conductivity using PCM

To determine the lateral resolution of PCM, we acquired images of regions containing platinum steps. We observed that the graphite layer covers the Pt steps continuously from the upper terrace to the lower adjacent one, as shown in Fig. 26. It is important to mention that the image in Fig. 26 is a contact current image. At distances far from the step in this image, the average current is the same on both terraces. However, close to the step, on what we have identified as the lower terrace, the contact current decreases by approximately 30%. At this scale, the topography image shows no contrast, since the graphite layer is almost flat,

Fig. 26. PCM image of a Pt step covered by a continuous layer of graphite. The topography image (not shown) is completely flat, and does not reveal the presence of a step in the graphite layer at this scale. Image size is 10 nm x 10 nm. The average current is 39 µA on the upper terrace and 28 µA on the lower terrace at a bias of 1.0 V.

although tilted with respect to the Pt substrate. On larger scale images of regions containing wide Pt terraces, it is possible to measure a height difference between the terraces. The same (5 x 5) moiré superstructure was detected on both sides of the Pt step, which indicates that a continuous sheet of graphite is covering the step.

In these experiments, we noticed that the tip-sample contact is not always conductive, unlike in previous experiments with similar cantilevers (Enachescu et al., 1998), possibly because of contamination as a result of gases used during sample preparation. In particular, when such contamination is observed, current vs. load curves indicate that the current is often not proportional to contact area, with a weak dependence on load, much less than would be expected from contact area variations. The step observed by PCM in Fig. 26 is about 1.5 nm wide, denoting a lateral resolution in PCM mode of this value. Using the DMT contact mechanics model as we did earlier, we estimate that the diameter of the contact area is approximately 8 nm, which indicates a contact AFM lateral resolution of no less than 8 nm. We can use the Sharvin model for point-contact resistance (Sharvin, 1965; Wexler, 1966) and the measured point-contact resistivity to estimate the area through which current flows in our contact. The diameter of this area is estimated to be 0.9 nm, which is consistent with the observed lateral resolution in the PCM image.

One explanation for our observation of the different resolutions in AFM and PCM is that the tip is covered with a poorly conducting layer, which is partially broken when the tip is cleaned by applying voltage pulses. This phenomenon may be limited to the tungsten-carbide coating of the tip used in this experiment. Another explanation may be that only the highest-pressure region of the contact area contributes to the point-contact current. However, a graph of current vs. load strongly favors the former explanation. We note that the weak dependence of current on load indicated by these graphs resembles the similarly weak I-Z dependence observed in the past in STM experiments on graphite (Salmeron et al., 1991). This supports the idea that, in most cases, STM on graphite is actually point-contact imaging. The change in local conductivity over the Pt step is likely due to the increased distance between the graphite layer and the underlying Pt substrate. The increased distance acts much like a tunneling barrier. In our measurements, we are able to measure current independently of topography, since the tip-sample contact is affected only by the mechanics of the system. The STM technique uses feedback on current to measure topography, so, for example, in the case of the blanketed Pt step, the STM tip would see the decrease in current and move closer to the sample to compensate. Thus, an STM image of a blanketed step would show a topographic step in the graphite layer with a width of 0.2 nm (i.e., typical STM resolution), while contact AFM indicates that the step width is many tens of nanometers. This width is the distance from the platinum step where the graphite layer begins to separate from the platinum substrate. Since the PCM technique is capable of separating mechanical and electrical measurements, it can offer additional insight into the electronic and tribological properties of surfaces.

The STM images of Land et al. (Land et al., 1992a; 1992b) indicate that there is local conductivity modulation at both the lattice and the moiré periodicities. If we imagine the atoms in our AFM contact contributing to the contact current as a collection of STM tips, one for each atom, the total contact current would be the sum of the contribution of these tips. We would still expect to see both the lattice and the moiré periodicities in the resulting PCM image, although the magnitude of modulation relative to the average current would decrease. The modulation would sum to zero only in special, destructively interfering cases. This will be discussed in more detail in a future work.

5. Sensing of dipole fields force in scanning tunneling and force microscopy experiments

The electric field of dipoles localized at the atomic steps of metal surfaces due to the Smoluchowski effect were measured from the electrostatic force exerted on the biased tip of a scanning tunneling microscope. By varying the tip-sample bias the contribution of the step dipole was separated from changes in the force due to van der Waals and polarization forces. Combined with electrostatic calculations, the method was used to determine the local dipole moment in steps of different heights on Au(111) and on the 2-fold surface of an Al-Ni-Co decagonal quasicrystal.

5.1 Background

The different electronic structure of the atoms at steps and terraces of metal surfaces is thought to be responsible for their different (often-enhanced) chemical reactivity. Dipole moments are postulated to exist localized at the steps due to incomplete screening of the positive ion cores by conduction electrons, because the spatial variation of the charge density is limited by the Fermi wavelength. This is known as the Smoluchowski effect (Smoluchowski, 1941). Indirect support for this assumption is provided by work function (φ) measurements. Besocke and Wagner found a decrease in φ proportional to the step density on Au(111) (Besocke & Wagner, 1973) and used this to estimate the average value of the step dipole. Similar results have been reported for Pt(111) and W(110) (Kral-Urban et al., 1977l; Besocke et al., 1977). Calculations using the jellium model (Ishida & Liebsch, 1992) predict that the localized step dipole increases with step height and screening length. Electronic structure calculations for the (111) and (100) microfacet steps on Al(111) produced very small dipole moments (Stumpf & Scheffler, 1996), indicating that the Smoluchowski effect alone is insufficient to fully describe the electronic structure of steps. It is therefore important that the presence and the magnitude of local dipole moments at steps be measured experimentally.

Scanning probe microscopy can be used to investigate the electronic structure of steps. Marchon et al. observed a reduction in the tunneling barrier at surface steps on sulfur-covered Re(0001) (Marchon et al., 1988) using scanning tunneling microscopy (STM). Later Jia et al. used this effect to calculate the step dipole for Au(111) and Cu(111) (Jia et al., 1998a; 1998b). Arai and Tomitori investigated step contrast as a function of tip bias on Si(111) (7x7) using dynamic atomic force microscopy (D-AFM) (Arai & Tomitori, 2000) and suggested that step dipoles could explain their observations. In contrast Guggisberg et al. investigated the same system using STM feedback combined with D-AFM force detection and concluded that the step dipole moments in Si(111)-(7x7) were negligible (Guggisberg et al., 2000). They attributed the D-AFM contrast effects to changes in the van der Waals and electrostatic polarization forces, which are reduced above and increased below the step edges relative to the flat terrace.

In this work we report measurements of the strength of the fields produced by the step dipoles through direct measurement of the electrostatic force they produce on biased tips. We use a combined STM-AFM system (Enachescu et al., 1998; Park et al., 2005b) with cantilevers that are made conductive by a ~30 nm coating of W_2C. Relatively stiff cantilevers of 48 or 88 N/m were used to avoid jump-to-contact instabilities close to the surface. Attractive forces cause the cantilevers to bend toward the surface during imaging, as illustrated in Fig 27(a). Scanning is done at constant current as in standard STM mode, while forces are measured simultaneously from the cantilever deflection (Park et al., 2005c).

The force acting on the tip is the sum of van der Waals and electrostatic contributions. The former is independent of the applied bias. The electrostatic contributions are additive and can be written as (Jackson, 1975):

$$F = f(D / R)V^2 + g(D / R)PV + h(D / R)P^2 \qquad (1)$$

where D is the tip-surface distance, R the tip radius and f, g and h are functions of the tip and sample geometry. P is the dipole moment, and V is the electrostatic potential difference between tip and sample. The first term in (1) represents the attractive force from polarization (i.e. image charges) induced by the applied voltage. The second term is due to surface dipoles P interacting with the biased tip, and is proportional to the bias. The last term is the force between the dipole P and its image on the tip. Of these contributions only the second term is linear with applied voltage, and provides an easy way to determine the net effect of the dipole field.

5.2 Experimental

The measurements were carried out in ultra high vacuum with an optical deflection AFM. Several samples were used, including Pt(111), Au(111) and the two-fold surface of a $Al_{74}Ni_{10}Co_{16}$ decagonal quasicrystal prepared by cutting the crystal parallel to the ten-fold axis. The growth and characterization of the Al-Ni-Co quasicrystal are outlined in detail elsewhere (Fisher et al., 1999). Due to the aperiodic nature of the atomic layering in the latter sample, steps of various heights were readily obtained on a single surface. The Pt single crystal and the quasicrystal (Park et al., 2004) samples were sputtered and annealed in UHV.

The Au sample was in the form of a thin film on glass, prepared in air by flame annealing and transferred to vacuum without further treatment. An average tip radius of 30-70 nm was determined by SEM imaging.

5.3 Results and discussions

Figure 27(b) and 27(d) shows the STM topography and force image of Pt(111) obtained simultaneously for a tip bias of –0.2 V. Fig 27(c) is a height and force profiles across the line in (b). The force, which is always attractive, increases by ~1.5 nN as the tip approaches the bottom of the step and decreases by ~4 nN after climbing over the step. When the attractive force increases, the STM current feedback loop retracts the base of the cantilever to keep the tunnel current, and hence the tip-sample distance, constant. The reduction of attractive force in the upper side of the steps is due to the reduction in the van der Waals and polarization part of the force (image charges), since in that position half of the surface (the lower terrace) is farther away from the tip. This is consistent with the results of Guggisberg (Guggisberg et al., 2000). By itself this result does not prove the existence of localized dipoles at the steps. For that we need to examine the changes in the force due to applied bias.

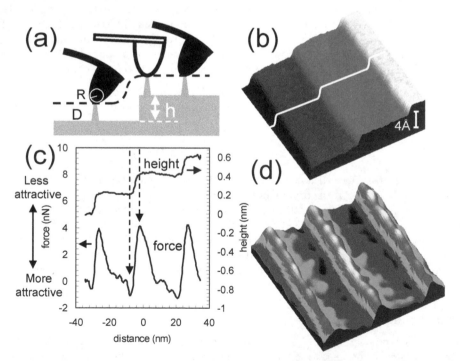

Fig. 27. (a) STM-AFM configuration using a conductive cantilever bending in response to forces. (b) 70 nm x 70 nm STM image of a Pt (111) surface (V_t = –0.2V, I=0.16nA). (c) Height and force profile across the steps. The force on the tip is more attractive at the bottom of the steps and less attractive at the top. (d) Force image simultaneously acquired with (b). Yellow and blue colors represent low and high attractive forces, respectively.

Earlier studies of decagonal Al-Ni-Co quasicrystal surfaces (Kishida et al., 2002) indicate that the bulk structure consists of pairs of layers with 5-fold quasiperiodic structure stacked along the 10-fold direction with a periodicity of 0.4 nm. In our 2-fold surface this produces rows of atoms arranged periodically. The rows are separated by distances varying in an aperiodic manner and are parallel to the step edges. Most steps have heights of 0.5, 0.8 and 1.3 nm, although a few are observed also with 0.2 nm. The ratios of these heights follow the golden mean (τ ~1.618), characteristic of their quasiperiodic nature. Fig. 28(a) shows a topographic profile perpendicular to the 10-fold axis, along with corresponding force profiles acquired at +1.2 and –1.2 V tip bias (at 100 pA tunneling current). Fig 28(b) shows similar topographic and force profiles across single and double-height steps on Au(111) at +3 and –3 V tip bias. Like in the Pt case, there is a reduction of the attractive force when the tip crosses over the steps (upward peaks in the force profile). While this reduction is present for both + and – bias, there is a noticeable difference between the two. The difference between forces at opposite biases eliminates all contributions except that from the second term in equation (1), which is purely due to the step dipole. We can immediately conclude that the positive end of the step dipole points up, consistent with a smaller attractive force at positive tip bias.

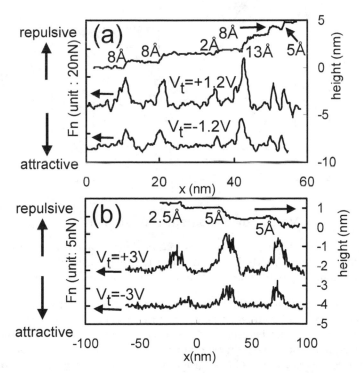

Fig. 28. (a) Height and force profiles across steps for positive and negatively biased tip (I = 0.1nA) on the Al-Ni-Co quasicrystal surface showing steps of multiple heights (0.2, 0.5, 0.8 and 1.3 nm). (b) Height and force profile across steps on a Au(111) surface. Small relative peak shifts in the force profiles are caused by noise and thermal drift. V_t is the tip voltage with respect to the sample.

Approach curves (force and current versus distance at fixed bias) were used to determine an STM tip-sample distance of 0.5 ± 0.1 nm during tunneling as shown in Fig. 29. Tunnel current vs. voltage curves for all samples showed a metallic character, with no significant dependence on bias polarity, so there is no change in the tip-sample distance under STM feedback when polarity is reversed. Force *vs* voltage curves over flat terraces reveal a small tip-sample contact potential difference of 0.14 V for the quasicrystal and 0.20 V for gold. This contact potential difference is negligible compared with the applied bias and cannot account for the polarity-dependent force contrast at step edges.

The tip radius can be extracted from the force-distance curves as described in previous work (Sacha et al., 2005) that shows that the effective tip radius is given by $R = 36A/V^2$, where A is the slope in the plot of electrostatic force F, versus $1/D$, F is in nanonewton, $1/D$ in nm^{-1}, V in volts, and R in nm, as shown in the inset of Fig. 29.

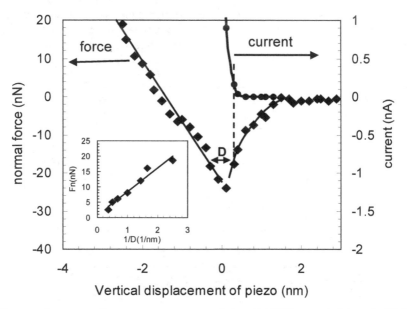

Fig. 29. Force and current-distance curves measured on Au(111) at a tip bias of –3V. Before contact the electrostatic force bends the tip towards the surface. This attraction is used to calculate the tip radius (inset), from the slope of F vs. $1/D$, yielding R =30 nm.

Results from measurements using the polarity-dependent component of the force (i.e., the difference between forces at V^+ and V^- bias, divided by $2|V|$) at steps of various heights are shown in figure 30(a). As can be seen, the experimental points follow a straight line. To determine the magnitude of the step dipole moment, we compute the electrostatic force using the *Generalized Image Charge Method* (GICM) program (Mesa et al., 1996; Gómez-Moñivas et al., 2001), a variational method for solving electrostatic problems that is particularly efficient for problems with high symmetry. The tip is modeled by a sphere of radius R, which is an equipotential surface produced by a series of point charges q_i and dipoles p_i at fixed positions r_j within the sphere. The magnitudes of the charges are adjusted to reproduce the boundary conditions of a constant potential V at radius R, and the sample surface at ground. With a suitable choice of positions, a relatively small number of point charges (less than 10) can reproduce the potential over the surface of the sphere within ~1%. In this method the relative positions of the point charges and dipoles within the tip are fixed; only the magnitudes of the charges are changed as the tip-sample geometry is changed.

For the present geometry, six point charges were distributed along the surface normal between the center and the sphere boundary, plus two symmetrical pairs of point dipoles located off-axis in the plane defined by the surface normal and the line dipole P. Once the effective charges were determined, the tip-sample forces were calculated as the sum of the forces between the point charges q_i, p_i, their image charges q_i', p_i' below the surface plane, and the fixed line dipole P. The field distribution calculated using these parameters is shown in Fig. 30(b).

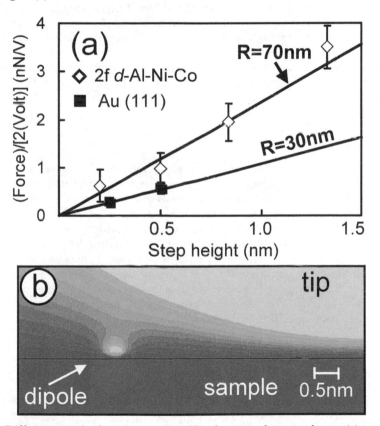

Fig. 30. (a) Difference in the force experienced by the tip at the steps for positive and negative bias, per unit applied volt [nN/V]. Open symbols correspond to steps on Al-Ni-Co quasicrystal surface. Filled symbols to steps on Au(111). The lines are calculations for 1 Debye and 0.45 Debye per step atom respectively; by definition they pass through the origin. The error bar is associated with the noise level of force measurement. (b) Electric field distribution calculated using the GICM in the tip-sample region with a permanent dipole close to the tip apex. (R=30 nm, D=0.5 nm).

The radii of the tips used for the Au and Al-Ni-Co samples derived from the force-distance curves was found to be 30 ± 11 nm, and 70 ± 30 nm respectively. Calculations performed for several values of tip radius and for 0.5 nm for D, are shown in figure 30(a) as a function of step height and step dipole moment (Park et al., 2005a). As we can see the data (difference in the force at + and – bias per unit applied volt) fit well the lines corresponding to step dipole values of 1.6 Debye/nm or 0.45 Debye/step atom for Au(111) monoatomic steps (Park et al., 2005a) with the tip radius of 30 nm, and 2.5 Debye/nm or 1.0 Debye/step atom for the smallest (0.2 nm) quasicrystal steps with the tip radius of 70 nm. We can conclude that the dipole moment scales proportionally to step height, at least for steps up to 1.5 nm.

The dipole moment obtained for Au(111) is ~3 times larger than the value of 0.16 Debye/atom obtained by Jia *et al.* (Jia et al., 1998a; 1998b) from STM barrier height measurements and ~2 times larger than the 0.20 to 0.27 Debye/atom obtained by Besocke *et al.* (Besocke & Wagner, 1973) from work-function measurements on stepped Au(111). Bartels *et al.* (Bartels et al., 2003) obtained 0.33 Debye/atom for Cu(111) steps from STM spectroscopy of localized states at step edges. Apart from systematic and statistical errors in the measurements, the discrepancy could be related to the very different methods used, tunneling barrier in one case and average work function in another as compared to direct measurement of the dipole force field in the present work.

6. Conclusions

In conclusion, we have studied the nanotribological behavior of a well-defined hydrogen-terminated diamond(111)/tungsten-carbide single asperity contact in UHV as a function of applied load. Local contact conductance measurements showed no significant changes in the shape of I-V curves for loads up to 1.7 µN, as expected from the proportionality between the current and the contact area, which provided us with a direct and independent way of measuring the area of contact. The DMT model provided an excellent fit to the current (contact area) versus load data for a variety of bias voltages, which is in agreement with the finding that $\mu = 0.02 < 0.1$. Using the DMT-relation for contact area versus load, we found that for this ideal single-asperity contact, *i.e.* one of the hardest, stiffest known heterocontacts, involving materials of great tribological importance, friction is directly proportional to the contact area: $F_f = \tau A$, where $\tau = 238$ MPa for loads up to 12 nN.

By using AFM and STM with the same conductive WC-coated cantilevers, we were able to study the tribological and electronic properties of nanocontacts and to correlate these properties with the degree of passivation of the interface. Contacts could be classified as clean, half-passivated and fully passivated, depending on whether none, one or both of the participating surfaces are covered with chemically inactive layers. While it would be desirable to obtain detailed information on the specific chemistry and structure of these contaminant species, no technique currently exists for obtaining such information at a confined nanoscale interface. Rather, we are restricted to rely on wide-scale AES measurements of the surfaces. Based on these measurements, we propose that the passivating materials for the WC tips consist mostly of strongly bound O and C species. On the tip they could be removed by sliding contact under high load on the Pt substrate. In the case of Pt, the contaminants were C species.

The clean Pt(111) surface could be imaged in STM mode with cantilevers stiff enough to avoid the jump-to-contact instability. When such a surface is brought into contact with a clean tip, strong bonds are formed that cause rupture of the contact in the bulk part of the tip and/or substrate upon separation. Sliding is strongly impeded in this case and always leads to severe cantilever deformations and distorted force-displacement curves.

With passivated tips, low adhesion energy contacts ($\sim$1 J/m^2) are formed. The friction properties of such contacts depend on whether additional adsorbate layers are also present on the Pt surface. Passivated areas of the surface give rise to low friction and sigmoid-type I-V characteristics, typical of poorly conductive or semiconducting materials. Clean Pt areas produce Ohmic contact characteristics.

Clean Pt can be imaged in contact mode with passivated tips and gives rise to atomic lattice stick-slip friction with the Pt(111) lattice periodicity. This is the first time that a chemically active metal surface has been imaged in UHV in AFM contact-mode, revealing stick-slip with atomic lattice periodicity, and indicates that the passivating layer on the WC tip is bound strongly enough to the tip that material is not transferred to the active Pt even in conditions where substantial energy dissipation takes place during friction.

The results indicate that even in ultrahigh vacuum conditions, transfer of low-conductivity, passivating material can easily occur in nano-scale contacts. This demonstrates that detailed studies of third-body processes at the nanoscale are accessible with this AFM-STM multifunctional approach. The presence of these species substantially effects friction and adhesion. These results are relevant to the understanding of transfer film formation and its influence on the structural evolution and tribology of interfaces, whose inelastic properties are only beginning to be probed and understood at the nanometer scale.

We presented the first results of the combination of PCM and AFM techniques, in which current images, obtained on contacts many nanometers in diameter produced by very high loads (up to 5 GPa), reveal the atomic scale periodicity of the substrate. This surprising observation indicates that, even after averaging over many contact points of atomic dimension, the lattice periodicity does not average out.

We also showed that PCM is capable of measuring variations in local conductivity with a lateral resolution that is similar to the corresponding AFM resolution. Moreover, the technique is capable of separating mechanical and electrical contributions to the measured current. We were able to determine that local conductivity variations arise from different sources, namely, moiré superstructure and the conductivity to the underlying substrate.

We favor point-contact current imaging of lattice resolution as an explanation for many of the STM images on graphite presented in the past, especially in the first decade of STM experiments. In these experiments, it is likely that the tip was in contact with the surface, as in PCM, which explains the weak dependence of "tunneling" current as a function of tip distance.

Point contact current imaging, in conjunction with simultaneous friction and topographic imaging, should be an important tool in current efforts to understand the atomic origin of friction. We are currently applying these techniques to study the tribological behavior of surfaces.

Finally, we have shown the existence of localized dipole fields in the vicinity of steps through direct measurements of the forces experienced by a biased STM tip. Together with measurements of the tip radius (from force-distance curves) and tip-sample distance (from current-distance approach curves) in the course of the same experiment, the method provides a direct way to map out and to measure local dipole moments on surfaces that should be of significance in studies of chemical and electronic properties of surfaces.

7. Acknowledgments

The work presented here was done at Materials Sciences Division, Lawrence Berkeley National Laboratory, Berkeley, CA 94720

Many thoughts, analyses, suggestions, and conclusions in this work were generated from research by and discussions with the following people: Dr. R.W. Carpick, Dr. D. F. Ogletree, Dr. J. Y. Park, Dr. R.J.A. van den Oetelaar, Dr. X. Lei, Mr. D. Schleef and Dr. M. Salmeron.

This work was supported by the Director, Office of Energy Research, Office of Basic Energy Sciences, Materials Sciences Division, of the U.S. Department of Energy under Contracts No. DE-AC03-76SF00098, DE-FG02-02ER46016, and DE-AC02-05CH11231. Also, this work was supported by the Ministry of Education, Research, Youth and Sport, Romania, and by the European Union through the European Regional Development Fund, and by Romanian National Authority for Scientific Research, under project POSCCE-O 2.1.2-2009-2/12689/717.

8. References

Arai, T. & Tomitori, M. (2000). Bias dependence of Si(111) 7x7 images observed by noncontact atomic force microscopy. *Appl. Surf. Sci.*, Vol. 157, pp. (207-211)

Atamny, F. & Baiker, A., (1995). Direct imaging of the tip shape by AFM. *Surf. Sci.*, Vol. 323, pp. (L314 –L318)

Barrena, E., Kopta, S., Ogletree, D.F., Charych, D.H., & Salmeron, M., (1999). Relationship between friction and molecular structure: Alkylsilane lubricant films under pressure. *Phys. Rev. Lett., Vol.* 82, pp. (2880-2883)

Bartels, L., Hla, S. W., Kühnle, A., Meyer G. & Rieder, K. H. & Manson, J. R. (2003). STM observations of a one-dimensional electronic edge state at steps on Cu(111). *Phys. Rev.B*, Vol. 67, No. 20, pp.(205416-205420)

Batra, I. P., Garcia, N., Rohrer, H., Salemink, H., Stoll, E. & Ciraci, S. (1987). A study of graphite surface with STM and electronic-structure calculations. *Surf. Sci.*, Vol. 181, pp. (126-138)

Batra I. P. & Ciraci, S. (1988). Theoretical scanning tunneling microscopy and atomic force microscopy study of graphite including tip surface interaction. *J. Vac. Sci. Technol.* A, Vol. 6, pp. (313-318)

Besocke, K. & Wagner, H. (1973). Adsorption of W on W(110): Work-Function Reduction and Island Formation. *Phys. Rev.B,* Vol. 8, pp. (4597-4600)

Besocke, K., Krahl-Urban B. & Wagner, H. (1977). Dipole moments associated with edge atoms; A comparative study on stepped Pt, Au and W surfaces. *Surf. Sci.*, Vol. 68, pp. (39-46)

Binnig, G., Rohrer, H., Gerber H., & Weibel E. (1983). 7x7 reconstruction on Si(111) resolved in real space. *Phys. Rev. Lett.*, Vol. 50, pp. (120-123)

Binnig, G., Quate, C.F. & Gerber, Ch. (1986). Atomic Force Microscope. *Phys. Rev. Lett.* Vol. 56, pp. (930- 933)

Binnig, G., Fuchs, H., Gerber, C., Rohrer, H., Stoll, E., & Tosatti, E. (1986). Energy-dependent state-density corrugation of a graphite surface as seen by scanning tunneling microscopy. *Europhys. Lett.*, Vol. 1, pp. (31-36)

Bennewitz, R., Gyalog, T., Guggisberg, M., Bammerlin, M., Meyer, E., & Güntherodt, H.J., (1999). Atomic-scale stick-slip processes on Cu(111). *Phys. Rev. B*, Vol. 60, pp. (R11301-R11304)

Bennewitz, R., Gnecco, E., Gyalog, T., & Meyer, E., (2001). Atomic friction studies on well-defined surfaces. *Trib. Lett.*, Vol. 10, pp.(51-56)

Dai, Q., Vollmer, R., Carpick, R.W., Ogletree, D.F., & Salmeron, M., (1995). Variable-temperature ultrahigh-vacuum atomic-force microscope . *Rev. Sci. Instrum.*, Vol. 66, pp. (5266-5271)

Daly, C., & Krim, J. (1996). Friction and damping of Xe/Ag(111). *Surf. Sci.*, Vol. 368, pp. (49-54)

Derjaguin , B. V., Muller, V. M., Toporov, Y. P., (1975). Effect of contact deformations on the adhesion of particles. *J. Colloid Interface Sci.*, Vol. 53, pp. (314-326)

Dowson, D. (1998). *History of Tribology*, Longman, London

Enachescu, M., van den Oetelaar R.J.A., Carpick R.W., Ogletree D.F., Flipse C.F.J., & Salmeron M., (1998). Atomic force microscopy study of an ideally hard contact: The diamond(111)/tungsten carbide interface. *Phys. Rev. Lett.*, Vol. 81, pp. (1877-1880)

Enachescu, M., van den Oetelaar, R.J.A., Carpick, R.W., Ogletree, D.F., Flipse C.F.J., & Salmeron, M., (1999). Observation of proportionality between friction and contact area at the nanometer scale. *Trib. Lett.*, Vol. 7, pp. (73-78)

Enachescu, M., Schleef, D., Ogletree, D.F. & Salmeron, M.(1999). Integration of Point-Contact Microscopy and Atomic-Force Microscopy: Application to Characterization of Graphite/Pt(111). Phys. Rev. B, Vol. 60, pp. (16913-16919)

Enachescu, M., Carpick, R.W., Ogletree, D.F., & Salmeron, M.(1999). Making, breaking and sliding of nanometer-scale contacts. *Mat. Res. Soc. Symp. Proc.*, Vol. 539, pp. (93-103)

Enachescu, M., Carpick, R.W., Ogletree, D.F., & Salmeron, M.(2004). The role of contaminants in the variation of adhesion, friction, and electrical conduction properties of carbide-coated scanning probe tips and Pt(111) in ultrahigh vacuum. *J. App. Phys.*, Vol. 95, No. 12, pp. (7694-7700)

Carpick, R.W., Agraït, N., Ogletree, D.F., & Salmeron, M. (1996). Measurement of interfacial shear (friction) with an ultrahigh vacuum atomic force microscope, *J. Vac. Sci. Technol. B*, Vol.14, pp. (1289-1295)

Carpick, R. W., Agraït, N., Ogletree, D. F., & Salmeron, M., (1996). Variation of the interfacial shear strength and adhesion of a nanometer-sized contact. *Langmuir*, Vol. 12, pp. (3334-3340)

Carpick, R.W. & Salmeron, M., (1997). Scratching the surface: Fundamental investigations of tribology with atomic force microscopy . *Chem. Rev.* Vol. 97, pp. (1163-1194)

Carpick R.W., Enachescu M., Ogletree D.F., & Salmeron M., (1998). Making, Breaking and Sliding of Nanometer-Scale Contacts, *Proceedings of the MRS Fall Meeting*, Boston, October 1998

Carpick, R.W., Dai, Q., Ogletree, D.F., & Salmeron, M., (1998). Friction force microscopy investigations of potassium halide surfaces in ultrahigh vacuum: structure, friction and surface modification .*Tribol. Lett.*, Vol. 5, pp. (91-102)

Cottrell A.H., (1988). *Introduction to the Modern Theory of Metals*, Institute of Metals, ISBN 0-904357-97-X, London

Drummond, C., Alcantar, N., Israelachvili, J., Tenne, R., Golan, Y., (2001). Microtribology and friction-induced material transfer in WS2 nanoparticle additives. *Advanced Functional Materials*, Vol. 11, pp.(348-354)

Fisher, I. R., Kramer, M. J., Islam, Z., Ross, A. R., Kracher, A., Wiener, T., Sailer, M. J., Goldman, A. I., & Canfield, P. C. (1999). On the growth of decagonal Al-Ni-Co quasicrystals from the ternary melt.*Philos. Mag. B*, Vol. 79, pp. (425-434)

Gardos, M.N., (1994). *Emerging CVD Science and Technology*, In *Synthetic Diamond*, K.E. Spear and J.P. Dismukes, pp. (419- 425), Wiley, New York

Gómez-Moñivas, S., Froufe-Pérez, L.S., Caamaño A.J. & Sáenz, J.J. (2001). Electrostatic forces between sharp tips and metallic and dielectric samples. *Appl. Phys. Lett.*, Vol. 79, pp. (4048-4050)

Greenwood, J.A., & Williamson, J.B.P. (1966). *Proc. R. Soc. London, Ser. A*, Vol. 295, pp. (300-309)

Greenwood, J.A., (1997). *Proc. R. Soc. London, Ser. A* Vol. 453, pp. (301-306)

Greenwood, J. A. & Johnson, K. L. (1998). An alternative to the Maugis model of adhesion between elastic spheres. *J. Phys. D*, Vol. 31, pp. (3279-3290)

Grierson, D.S. & Carpick R.W. (2007). Nanotribology of Carbon-based Materials. *Nano Today*, Vol. 2, No. 5, pp. (12-21)

Guggisberg, M., Bammerlin, M., Baratof, A., Lüthi, R., Lopacher, Ch., Battison, F. M., Lü, J., Bennewitz, R., Meyer, E. & Güntherodt, H.-J. (2000). Dynamic force microscopy across steps on the Si(111)-(7 x 7) surface. *Surf. Sci.*, Vol. 461, pp. (255-265)

Harrison, J.A., & Brenner, D.W. (1995). *Handbook of Micro/Nanotribology*, CRC Press, ISBN-10: 084938401X, Boca Raton

Hansma, P. K. (1985). *Bull. Am. Phys. Soc.*, Vol. 30, pp. (251-254)

He, G., Muser, M. H., & Robbins, M. O., (1999). Adsorbed layers and the origin of static friction. *Science*, Vol. 284, pp. (1650-1652)

He, G. & Robbins, M. O., (2001). Simulations of the static friction due to adsorbed molecules. *Physical Review* B, Vol. 64, No. 3, pp. (035413-035425)

Homola, A.M., Israelachvili, J.N., Gee M.L., & McGuiggan P.M., (1989). Measurements of and relation between the adhesion and friction of 2 surfaces separated by molecularly thin liquid-films. *J. Tribol.* Vol. 111, pp. (675-682)

Hu, Z., Ogletree, D. F., Van Hove, M. A. & Somorjai, G. A. (1987). Leed theory for incommensurate overlayers - application to graphite on Pt(111). *Surf. Sci.*, Vol. 180, pp. (433-459)

Hu, J., Xiao, X.-D., Ogletree, D.F., & Salmeron, M., (1995). Atomic-scale friction and wear of mica. *Surf. Sci.*, Vol. 327, pp. (358-370)

Hurtado, J. & Kim, K.-S., (1998) Fracture and Ductile versus Brittle Behavior: Theory, Modeling, and Experiment, *Proceedings of the MRS Fall Meeting*, Boston, 1998

Ishida, H. & Liebsch, A. (1992). Calculation of the electronic-structure of stepped metal-surfaces. *Phys. Rev. B*, Vol. 46, pp. (7153-7156)

Israelachvili, J.N. & Tabor, D. (1972). The Measurement of Van Der Waals Dispersion Forces in the Range 1.5 to 130 nm. *Proc. Roy. Soc. A*, London, November 1972, Vol.331, No. 1584, pp. (19-38)

Israelachvili, J.N., McGuiggan, P.M., & Gee, M.L., (1990). Fundamental experimental studies in tribology - the transition from interfacial friction of undamaged molecularly smooth surfaces to normal friction with wear. *Wear*, Vol. 136, pp. (65-83)

Israelachvili, J. N.(1992). *Intermolecular and Surface Forces*, Academic Press London, London

Jackson, J. D. (1975). *Classical Electrodynamics*, John Wiley & Sons, ISBN 978-0-471-43132-9, New York

Jansen, A.G.M., van Gelder, A.P., & Wyder, P., (1980). Point-contact spectroscopy in metals. *J. Phys. C (Sol. State Phys.)*. Vol.13, pp. (6073-6118)

Jia, J. F., Inoue, K. , Hasegawa, Y., Yang W. S. & Sakurai, T. (1998). Variation of the local work function at steps on metal surfaces studied with STM. *Phys. Rev. B*, Vol. 58, pp. (1193-1196);

Jia, J. F., Hasegawa, Y., Inoue, K., Yang, W. S. & Sakurai, T.(1998). Steps on the Au/Cu(111) surface studied by local work function measurement with STM. *Appl. Phys. A*, Vol. 66, pp. (S1125-S1128)

Johnson, K. L., Kendall, K., & Roberts, A. D., (1971). *Proc. Roy. Soc. London*, Vol. A 324, pp. (301-306)

Johnson, K.L., (1987). *Contact Mechanics*, Cambridge University Press, Cambridge

Johnson, K.L., (1996). Continuum mechanics modeling of adhesion and friction. *Langmuir* Vol.12, pp. (4510-4513)

Johnson, K.L., (1997). *Proc. R. Soc. London, Ser. A 453*, pp. (163-171)

Kishida, M., Kamimura, Y., Tamura, R., Edagawa, K., Takeuchi, S., Sato, T., Yokoyama, Y., Guo, J. Q. & Tsai, A. P. (2002). Scanning tunneling microscopy of an Al-Ni-Co decagonal quasicrystal. *Phys. Rev. B*, Vol. 65, No. 9, pp.(094208-094215)

Klein, C.A., (1992). Anisotropy of young modulus and poisson ratio in diamond. *Mater. Res. Bull.*, Vol. 27, pp. (1407-1414)

Kral-Urban, B., Niekisch, E. A & Wagner, H. (1977). Work function of stepped tungsten single crystal surfaces. *Surf. Sci.*, Vol. 64, pp. (52-68)

Krim, J., Watts, E.T. & Digel, J. (1990). Slippage of simple liquid-films adsorbed on silver and gold substrates. *J. Vac. Sci. Technol. A*, Vol. 8, pp. (3417-3420)

Krim, J., Solina, D.H., & Chiarello, R. (1991). Nanotribology of a Kr monolayer - a quartz-crystal microbalance study of atomic-scale friction. *Phys. Rev. Lett.*, Vol. 66, pp. (181-184)

Krim, J. (1996). Friction at the atomic scale. *Sci. Am.*, Vol. 275, pp. (74-80)

Kuwabara, M., Clarke, D. R. & Smith, D. A. (1990). Anomalous superperiodicity in scanning tunneling microscope images of graphite. *Appl. Phys. Lett.*, Vol. 56, pp. (2396-2398)

Land, T.A., Michely, T., Behm, R. J., Hemminger, J. C. & Comsa, G. (1992). STM investigation of single layer graphite structures produced on Pt(111) by hydrocarbon decomposition. *Surf. Sci.*, Vol. 264, pp. (261-270)

Land, T.A., Michely, T., Behm, R. J., Hemminger, J. C. & Comsa, G. (1992). Direct observation of surface-reactions by scanning tunneling microscopy - ethylene-]ethylidyne-]carbon particles-]graphite on pt(111). *J. Chem. Phys.*, Vol. 97, pp. (6774-6783)

Lantz, M.A., O'Shea, S.J., Welland, M.E., & Johnson, K.L. (1997). Atomic-force-microscope study of contact area and friction on NbSe2. *Phys. Rev. B*, Vol. 55, pp. (10776-10785);

Lantz, M.A., O'Shea, S.J., & Welland, M.E., (1997). Simultaneous force and conduction measurements in atomic force microscopy. *Phys. Rev. B*, Vol. 56, pp. (15345-15352)

Mamin, H. J., Ganz, E., Abraham, D. W., Thomson, R. E. & Clarke, J. (1986). Contamination-mediated deformation of graphite by the scanning tunneling microscope. *Phys. Rev. B*, Vol. 34, pp. (9015-9018)

Marchon, B., Ogletree, D. F. & Salmeron, M. (1988). Scanning tunneling microscopy study of the RE(0001) surface passivated by one-half a monolayer of sulfur in an atmospheric-environment. *J. Vac Sci. Tech. A*, Vol. 6, pp. (531-533)

Marti, O., Drake, B. & Hansma, P. K. (1987). Atomic force microscopy of liquid-covered surfaces - atomic resolution images. *Appl. Phys. Lett.*, Vol. 51, pp. (484-486)

Mate, C.M., McClelland, G.M., Erlandsson, R., & Chiang, S. (1987). Atomic-scale friction of a tungsten tip on a graphite surface. *Phys. Rev. Lett.* Vol. 59, pp. (1942- 1945)

Maugis D., (1992). Adhesion of spheres – The JKR-DMT transition using a Dugdale model. *J. Colloid Interface Sci.* Vol. 150, pp. (243-268)

McFadden, C.F. & Gellman, A., (1997). Metallic friction: the influence of atomic adsorbates at submonolayer coverages.*J., Surf. Sci.*, Vol. 391, pp. (287-299)

Mesa, G., Dobado-Fuentes E. & Sáenz, J. J. (1996). Image charge method for electrostatic calculations in field-emission diodes. *J. Appl. Phys.*, Vol. 79, pp.(39-44)

Müller, V.M., Yushenko, V.S., & Derjaguin, B.V., (1980). On the influence of molecular forces on the deformation of an elastic sphere and its sticking to a rigid plane. *J. Colloid Interface Sci.* Vol.77, pp. (91-101)

Müller, V.M., Derjaguin, B.V., & Toporov, Y.P., (1983). On 2 methods of calculation of the force of sticking of an elastic sphere to a rigid plane. *Colloids Surf.* Vol. 7, pp. (251-259)

Ogletree, D. F., Carpick, R. W., & Salmeron, M., (1996). Calibration of frictional forces in atomic force microscopy. *Rev. Sci. Instrum.*, Vol. 67, pp. (3298-3306)

Park, S.-I. & Quate, C. F. (1986). Tunneling microscopy of graphite in air. *Appl. Phys. Lett.*, Vol. 48, pp. (112-114)

Park, J. Y., Ogletree, D. F., Salmeron, M., Jenks C. J. & Thiel, P. A. (2004). Friction and adhesion properties of clean and oxidized Al-Ni-Co decagonal quasicrystals: a UHV atomic force microscopy/scanning tunneling microscopy study. *Tribol. Let.*, Vol. 17, pp. (629-636)

Park, J.Y., Sacha, G.M., Enachescu, M., Ogletree, D.F., Ribeiro, R.A., Canfield, P.C., Jenks, C.J., Thiel, P.A., Sáenz J.J., & Salmeron M. (2005). Sensing dipole fields at atomic steps with combined scanning tunneling and force microscopy. *Phys. Rev. Lett.*, Vol. 95, No. 13, 136802

(Park et al., 2005b) Park, J. Y., Ogletree, D. F., Salmeron, M., Ribeiro, R.A., Canfield, P.C., Jenks C. J. & Thiel, P. A.. (2005). Elastic and inelastic deformations of ethylene-passivated tenfold decagonal Al-Ni-Co quasicrystal surfaces. *Phys. Rev. B*, Vol. 71, No. 14, pp. (144203-144203-5);

Park, J. Y., Phaneuf, R.J., Ogletree, D.F., & Salmeron, M. (2005). Direct measurement of forces during scanning tunneling microscopy imaging of silicon pn junctions. *Appl. Phys. Lett.*, Vol. 86, No. 17, A pp. (172105-172105-3)

Persson, B. N. J., & Volokitin, A. I. (1995). Electronic friction of physisorbed molecules. *J. Chem. Phys.*, Vol. 103, pp. (8679-8683).

Persson, B. N. J. & Nitzan, A.(1996). Linear sliding friction: on the origin of the microscopic friction for Xe on silver. *Surf. Sci.*, Vol. 367, pp. (261- 275)

Qian, L. M., Xiao, X. D., & Wen, S. Z., (2000). Tip in situ chemical modification and its effects on tribological measurements. *Langmuir*, Vol. 16, pp. (662-670)

Rubio, G., Agraït, N., & Vieira, S., (1996). Atomic-sized metallic contacts: Mechanical properties and electronic transport . *Phys. Rev. Lett.*, Vol. 76, pp. (2303 -2305)

Sacha, G. M. et al. (2005). Effective tip radius in electrostatic force microscopy. *Appl. Phys. Lett.*, Vol. 86, No. 12, pp.(123101-1 -123101-3)

Salmeron, M. , Ogletree, D. F., Ocal, C., Wang, H. C., Neubauer, G., Kolbe, W. & Meyers, G. (1991). Tip surface forces during imaging by scanning tunneling microscopy. *J. Vac. Sci. Technol.* B, Vol. 9, pp. (1347-1352)

Schwartz, M.M., (1990). *Ceramic Joining*, Materials Park, June 1990).

Seki, H., McClelland, G.M., & Bullock, D.C.(1987). Raman-spectroscopy of disk coatings in a working magnetic disk drive. *Wear* Vol. 116, pp. (381-391)

Selloni, A., Carnevali, P., Tosatti, E., & Chen, C. D. (1985). Voltage dependent Scanning Tunneling Microscopy of a crystal surface graphite. *Phys. Rev.* B. Vol. 31, pp. (2602-2605)

Selloni, A., Carnevali, P , Tosatti, E., & Chen, C. D. (1986). Correction. *Phys. Rev.* B, Vol. 34, pp. (7406-7406)

Shackelford, J.F., Alexander, W., & Park, J.S. (1994) *CRC Materials Science and Engineering Handbook*, CRC Press, Boca Raton

Sharvin, Y. V. (1965). *Sov. Phys.*, JETP, Vol. 21, pp. (655-658)

Sheiko, S.S., Möller, M., Reuvekamp, E.M.C.M., & Zandbergen, H.W., (1993). Calibration and evaluation of scanning-force-microscopy probes. *Phys. Rev.* B, Vol. 48, pp. (5675-5678)

Singer, I. L. (1992). *Fundamentals of Friction: Macroscopic and Microscopic Processes*, I. L. Singer & H. M. Pollock, Vol. 220, pp. (237-245), Kluwer, Dordrecht.

Singer, I. L., (1998). How third-body processes affect friction and wear. *MRS Bulletin*, Vol. 23, pp. (37-40)

Smith, D. P. E., Binnig, G., & Quate, C. F. (1986). Atomic point-contact imaging. *Appl. Phys. Lett.*, Vol. 49, pp. (1166-1168)

Smoluchowski, R. (1941). Anisotropy of the Electronic Work Function of Metals. *Phys. Rev.*, Vol. 60, pp. (661-674)

Sokoloff, J. B.(1993). Fundamental mechanisms for energy dissipation at small solid sliding surfaces. *Wear*, Vol. 167, pp. (59- 68)

Sokoloff, J. B. (1995). Theory of the contribution to sliding friction from electronic excitations in the microbalance experiment. *Phys. Rev. B*, Vol. 52, pp. (5318-5322)

Soler, J. M., Baro, A. M., Garcia, N. & Rohrer, H. (1986). Interatomic forces in scanning tunneling microscopy - giant corrugations of the graphite surface. *Phys. Rev. Lett.*, Vol. 57, pp. (444-447)

Sridhar, I., Johnson, K. L. & Fleck, N. A. (1997). Adhesion mechanics of the surface force apparatus. *J. Phys.* D, Vol. 30, pp. (1710-1719)

Stumpf, R. & Scheffler, M. (1996). Ab initio calculations of energies and self-diffusion on flat and stepped surfaces of Al and their implications on crystal growth. *Phys. Rev.B*, Vol. 53, pp. (4958-4973)

Sze, S.M., (1981). *Physics of Semiconductor Devices*, J. Wiley & Sons, New York

Szlufarska, I, Chandross, M., & Carpick R.W. (2008). Recent advances in single-asperity nanotribology. *J. Phys. D: Appl. Phys.*, Vol. 41, pp. (123001-1 - 123001-39)

Tabor, D. (1977). Surface forces and surface interactions. *J. Colloid Interface Sci.* Vol.58, pp. (2-13)

Tersoff, J. (1986). Anomalous corrugations in scanning tunneling microscopy - imaging of individual states. *Phys. Rev. Lett.*, Vol. 57, pp. (440-443)

Tsai, H. & Bogy, D.B., (1987). Characterization of diamond-like carbon-films and their application as overcoats on thin-film media for magnetic recording . *J. Vac. Sci. Technol. A*, Vol. 5, pp. (3287-3312)

van den Oetelaar, R.J.A. & Flipse, C.F.J., (1997). Atomic-scale friction on diamond(111) studied by ultra-high vacuum atomic force microscopy. *Surf. Sci.*, Vol.384, pp. (L828-L835)

Yanson, I. K., Kulik, I. O. & Batrak, A. G. (1981). Point-contact spectroscopy of electron-phonon interaction in normal-metal single-crystals. *J. Low Temp. Phys.*, Vol. 42, pp. (527-556)

Watts, E.T., Krim, J., & Widom, A. (1990). Experimental-observation of interfacial slippage at the boundary of molecularly thin-films with gold substrates. *Phys. Rev. B*, Vol. 41, pp. (3466-3472)

Wexler, G. (1966). *Proc. Phys. Soc. London*, Vol. 89, pp. (927-930)

8

Predicting Macroscale Effects Through Nanoscale Features

Victor J. Bellitto[1] and Mikhail I. Melnik[2]
[1]Naval Surface Warfare Center
[2]School of Engineering Technology,
Southern Polytechnic State University, Marietta, GA
USA

1. Intoduction

Atomic force microscopy is extremely useful in the study of surface defects in crystals by providing topographical data at the nanometric scale. With the aide of advanced statistical analysis, nanoscale surface data acquired through atomic force microscopy can also be utilized to predict behavior at the macroscale. The behavioral model presented is the measure of shock sensitivity required to produce detonation of explosive crystal test samples. The surfaces studied were of 7 different varieties of (RDX) crystalline explosives from 5 manufacturers (Doherty & Watts, 2008). It has been speculated that particle size, crystal defects, density and crystal morphology may play a role in the shock sensitivity of RDX and there have been numerous attempts to quantify and/or link particular features of the explosive particles to the shock sensitivity behavior of their larger compositions (Doherty and Watts, 2008). The shock sensitivity data were obtained from model test compositions prepared as polymer-bonded explosives using hydroxy-terminated polybutadiene (HTPB) as the binder. The shock sensitivity, measured in a gap test, is the shock required to produce a detonation of the test composition 50% of the time. Varied card thicknesses of poly(methyl-methacrylate) (PMMA) are used to attenuate the initiating charge entering the sample tube. The shock pressure (GPa) impacting the sample is determined by the number of cards. A small number of cards translate to a larger shock and thus a less shock sensitive sample.

2. Experimental

The AFM analysis of the RDX crystal surfaces was performed using a Multimode V scanning probe microscope (Veeco Metrology Group). The instrument was operated in Tapping Mode, where topographical analysis is performed with minimal contact of the surface. The crystal topography is mapped by lightly tapping the surface with an oscillating probe tip. The sample surface topography modifies the cantilever's oscillation amplitude and the topography image is obtained by monitoring these changes while closing the z feedback loop to minimize them. A first order algorithm supplied by Veeco was used to "flatten" the images. The flatten command modifies the scanned image removing tilt and thus leveling the image.

A variety of surface features were observed including edge and screw dislocations, voids, cracks, peaks, valleys, plateaus, etc. Examples of images acquired of the RDX particles surfaces are shown in Fig. 1-7. They are presented as amplitude images since they more easily display the shape of the sample surface. The amplitude image is equivalent to a map of the slope of the sample. The z-scale shows the tip deflection as it encountered sample topography. The amplitude image on harder samples better highlights the edges of features while on softer samples it can depict subsurface features better than the topography image.

The root mean square (*RMS*) calculation (R) of the surface imagery acquired in height mode was used to determine the roughness and to quantify the different surface topologies. The roughness was calculated by finding the median surface height for the scanned image and then evaluating the standard deviation. The equation for determining the surface roughness is

$$R = \left(\frac{1}{MN} \sum_{k=0}^{M-1} \sum_{l=0}^{N-1} \left[z(x_k, y_l) - \mu \right]^2 \right)^{0.5},$$

where μ is the mean value of the height, z, across in-plane coordinates *(x,y)*:

$$\mu = \frac{1}{MN} \sum_{k=0}^{M-1} \sum_{l=0}^{N-1} z(x_k, y_l).$$

The necessity to add objectivity to the consistency across the surface is demonstrated by Fig. 1-7. Although the side by side images obtained are from the same particle they demonstrate two very different surface morphologies and/or different roughness measurements.

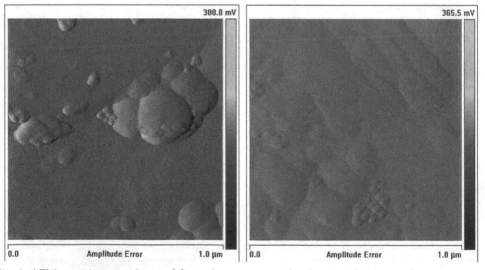

Fig. 1. AFM scan images obtained from the same particle of material I. A roughness measurement of 18.9 nm was obtained for image on the left while 29.6 nm was obtained for image on the right.

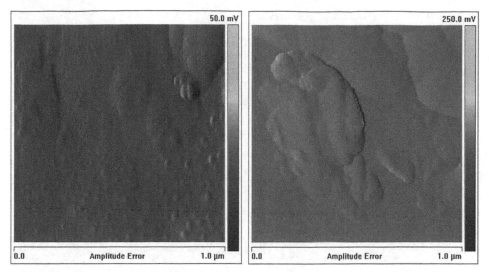

Fig. 2. AFM scan images obtained from the same particle of material II. A roughness measurement of 9.81 nm was obtained for image on the left while 24.4 nm was obtained for image on the right.

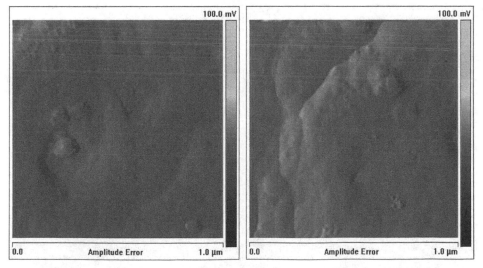

Fig. 3. AFM scan images obtained from the same particle of material III. A roughness measurement of 9.31 nm was obtained for image on the left while 21.5 nm was obtained for image on the right.

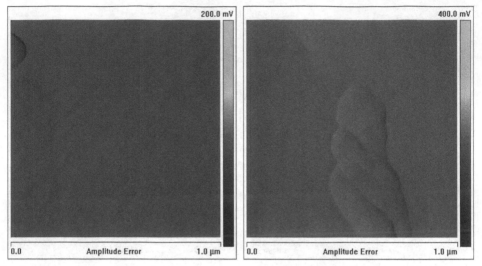

Fig. 4. AFM scan images obtained from the same particle of material IV. A roughness measurement of 5.05 nm was obtained for image on the left while 20.9 nm was obtained for image on the right.

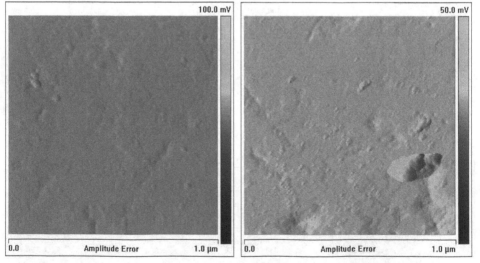

Fig. 5. AFM scan images obtained from the same particle of material V. A roughness measurement of 1.07 nm was obtained for image on the left while 3.7 nm was obtained for image on the right.

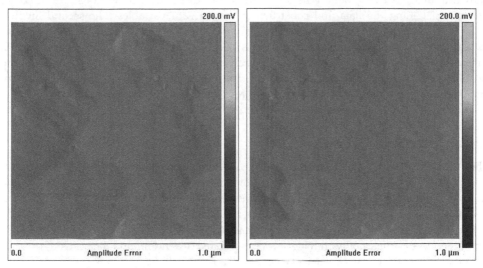

Fig. 6. AFM scan images obtained from the same particle of material VI. A roughness measurement of 1.07 nm was obtained for image on the left while 3.7 nm was obtained for image on the right.

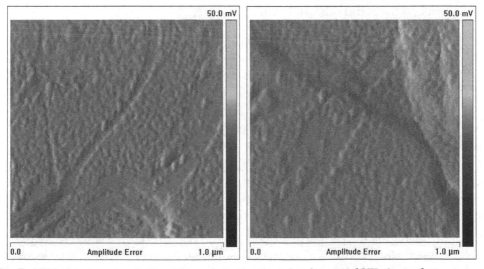

Fig. 7. AFM scan images obtained from the same particle of material VII. A roughness measurement of 2.9 nm was obtained for image on the left while 4.69 nm was obtained for image on the right.

3. Statistical analysis

Statistical analysis can generally be subdivided into three steps or sections. The first step involves obtaining the data and constructing the relevant variables. The second involves some basic statistical investigations, such as correlations between the variables or in the case of multiple sample comparisons, hypothesis testing. Often the first two steps may be sufficient, but in the event that the research requires an understanding of the exact relationships between the various variables, the quantitative effects one variable has on another, a further step in the form of a regression analysis becomes necessary. The discussion below outlines the steps encompassed in the statistical analysis.

3.1 Data and the construction of variables

The AFM data consist of individual 1 μm^2 surface scans obtained from seven different RDX materials. Each scan produces one observation that contains the information on the actual surface variation, roughness of the area of the scan, R (RMS). The scans were taken from seven different materials, however, each material is comprised of particles, and differences between particles of the same material in terms of their surface characteristics are possible. Therefore, five particles of each material were selected at random and 5-6 scans of different regions of the surface of each particle were acquired.

This approach allows modeling of the surface heterogeneity between the seven materials in three possible ways: the average measure of observed surface roughness, the variability between the particles, and lastly the variation in the surface roughness across the surface of a particle. For further discussion see Bellitto & Melnik (2010) and Bellitto et al (2010).

Multiple scans per particle allow the construction of two measures of the surface characteristics of the particle, the particle average (R_{pm}),

$$R_{pm} = \frac{1}{n}\sum_{s=1}^{n} R_{pms}$$

and the particle standard deviation (S_{pm}),

$$S_{pm} = \left(\frac{1}{n-1}\sum_{s=1}^{n}(R_{pms} - R_{pm})^2\right)^{0.5},$$

where the subscript p refers to the particle and m to the material. The subscript s refers to the individual scan and n is the number of scans per particle. Thus, R_{pm} represents the average value of all scans for that particle and S_{pm} represents the standard deviation of roughness for that particle. These statistics are listed in Table 1.

Shown in Table 1 are the particle average (R_{pm}) and the particle standard deviation (S_{pm}) values. These are limited to describing the individual particle characteristics but they can be used to construct variables describing the material characteristics. R_{pm} can simply be averaged to construct the average measure of observed surface roughness of the material (R_m),

Material	Particle	Particle Level Data		Material Level Data		
		Rpm	Spm	Rm	Sm	Smµ
I	1	19.100	6.531	18.270	4.131	5.929
	2	14.617	1.994			
	3	10.128	3.229			
	4	23.467	3.704			
	5	24.040	5.198			
II	1	10.518	7.730	8.425	4.853	6.120
	2	18.268	7.614			
	3	5.062	4.247			
	4	3.653	2.417			
	5	4.624	2.255			
III	1	13.794	8.236	16.040	11.033	7.457
	2	7.014	3.723			
	3	25.006	16.742			
	4	22.335	17.346			
	5	12.052	9.121			
IV	1	11.255	12.770	10.565	9.807	4.310
	2	4.872	3.160			
	3	10.220	11.350			
	4	16.912	17.749			
	5	9.564	4.007			
V	1	2.000	1.619	5.322	3.622	3.165
	2	6.207	1.698			
	3	10.244	7.900			
	4	3.396	4.034			
	5	4.762	2.856			
VI	1	15.740	4.633	10.067	3.612	4.784
	2	13.982	3.868			
	3	7.934	2.370			
	4	8.753	5.830			
	5	3.923	1.358			
VII	1	10.335	3.462	6.150	2.877	5.278
	2	5.120	2.169			
	3	12.820	7.432			
	4	1.482	1.041			
	5	0.994	0.281			

Table 1. Construction of Measures of Surface Roughness

$$R_m = \frac{1}{N} \sum_{p=1}^{N} R_{pm} \ ,$$

where N represents the number of particles for the material. Similarly, the average measure of particle standard deviation can be constructed,

$$S_m = \frac{1}{N} \sum_{p=1}^{N} S_{pm} \ .$$

This simple average measure accounts for the average variability in the surface roughness across the particle surface.

At this point two measures of surface roughness and its variability have been constructed. One is the particle average level for the material (R_m) and the other is the average variation in surface roughness across the particle surface (S_m). To quantify the variation between particles the standard deviation of the distribution of R_{pm}, is introduced and expressed as

$$S_{m\mu} = \left(\frac{1}{N-1} \sum_{p=1}^{N} (R_{pm} - R_m)^2 \right)^{0.5} \ .$$

This measure enables us to account for the heterogeneity between the various particles of the material.

Cyclotetramethylene-tetranitramine (HMX), a major impurity within RDX, has been reported to be as high as 17% (Doherty & Watts, 2008). Table 2 provides a basic summary of the shock sensitivity of the seven materials and the mean % of HMX impurity. The HMX impurity is included since impurities can significantly alter the periodicity of a crystal and thus affect its surface roughness.

Material	Impurity (Mean % HMX)	Sensitivity (GPa)
I	7.36	4.2
II	0.02	4.66
III	0.03	2.21
IV	8.55	3.86
V	0.82	5.24
VI	0.02	5.21
VII	0.19	5.06

Table 2. Impurity (%HMX) and shock sensitivity (GPa) of the materials used in our study.

3.2 Basic comparison of the materials

The average surface roughnesses of the materials, shown in Table 1, are first compared using analysis of variance (ANOVA). ANOVA is a statistical method used to compare population characteristics for multiple populations. ANOVA relies on three important assumptions, randomness and independence, normality, and homogeneity of variances. As discussed previously, our process of particle selection met the assumption of randomness. The Shapiro-

Wilk test is employed to test for normality, to test if the underlying population can be assumed to be normally distributed. Table 3 presents the results of the test for the particle (R_{pm}) values for all seven samples (materials). The Shapiro-Wilk test fails to reject the null hypothesis of normality with p=0.10 for all of the samples. The null hypothesis assumes that the underlying population is normally distributed. If the test fails to reject the null hypothesis then that indicates that the assumption of the underlying population being normally distributed cannot be disproved. However, the Shapiro-Wilk test may at times be misleading and a visual examination of the data may be recommended. Interestingly enough, the F test used in ANOVA is relatively robust against the assumption of normality (Levine et al, 2010), but the assumption of homogeneity of variances is crucial to the validity of the test.

Material	W	V	z	Prob>z
I	0.921	0.929	-0.097	0.538
II	0.828	2.032	1.107	0.134
III	0.937	0.739	-0.379	0.648
IV	0.952	0.565	-0.682	0.752
V	0.946	0.634	-0.556	0.711
VI	0.948	0.617	-0.585	0.721
VII	0.897	1.218	0.273	0.392

Table 3. Results of Shapiro-Wilk test

The Levene test for homogeneity of variances is performed on the R_{pm} data and the groups are defined as the individual materials. The test compares multiple samples to determine if they are drawn from populations with equal variances. The value of the F statistic from the Levene test is 0.49367, while the critical value for the rejection of the null hypothesis of homogeneity of variances is 2.445259, the hypothesis of homogeneity of variances cannot be rejected by the test[1]. This enables us to perform the one way ANOVA on our R_{pm} data, grouped by their corresponding materials.

The ANOVA method enables us to check if there is enough statistical evidence to reject the hypothesis that the R_m values of the seven materials are statistically not different from each other. The ANOVA output is presented in Table 4 and it shows that the hypothesis of equal R_m values is rejected by the data.

The ANOVA results demonstrate that these materials differ substantially in terms of their particle average roughness.

The focus of our research is to investigate a possible connection between the surface roughness and the shock sensitivity of the material. One simple way in which this connection can be examined is with the help of correlations, see Table 5. A clear negative correlation is observed between the measure of shock sensitivity (GPa) and the three

---

[1] In the event the Levene test rejected the null hypothesis, we would not be able to proceed with ANOVA and would have to use weaker testing techniques such as the Kruskal-Wallis test. For further discussion on the Levene test please see Levine et al (2010).

SUMMARY				
Groups	Count	Sum	Average	Variance
I	5	91.352	18.2704	35.15459
II	5	42.12567	8.425133	37.45042
III	5	80.20038	16.04008	55.60776
IV	5	52.823	10.5646	18.57295
V	5	26.60883	5.321767	10.01659
VI	5	50.33267	10.06653	22.88255
VII	5	30.7508	6.15016	27.86016

ANOVA						
Source of Variation	SS	df	MS	F	P-value	F crit
Between Groups	705.2395	6	117.5399	3.964342	0.005398	2.445259
Within Groups	830.1801	28	29.64929			
Total	1535.42	34				

Table 4. ANOVA of Rpm

	%HMX	Sensitivity	Rm	Sm
Sensitivity	-0.181			
	-0.411			
Rm	0.448	-0.694		
	1.120	-2.155		
Sm	0.284	-0.896	0.452	
	0.662	-4.518	1.133	
Smμ	-0.170	-0.706	0.656	0.414
	-0.386	-2.230	1.945	1.016

Table 5. Correlations and their statistical significance

measures of surface roughness of the material. This demonstrates that higher levels of surface heterogeneity are associated with lower levels of shock sensitivity.

Furthermore, all of the correlation coefficients between the surface characteristics measures and the shock sensitivity are statistically significant at or above the 90% level of significance. The statistical significance of the coefficients is determined by their corresponding t values. The t statistics for the significance test of the correlation coefficients are reported in italics under their corresponding coefficient and those that are significant at or above the 90% level are highlighted in bold font. The test statistic is obtained by the equation

$$ t = r\sqrt{\frac{N_m - 2}{1 - r^2}} \; , $$

where r is the correlation coefficient and N_m is the number of observations, which in this case is limited to seven, the number of materials used in this study.

3.3 Regression analysis

Regression analysis is designed to establish numerical relationships between the regressors, the independent variables and the dependent variable. A multivariable regression enables one to interpret the regression coefficients as partial derivatives of the dependent variable with respect to the regressor. Various regression techniques exist, but given the simple setup of our problem the most basic model, the Ordinary Least Squares, can adequately serve the purpose. For further discussion of various regression techniques see Greene (2003). The greater concern is the fact that the data is limited to only seven materials and seven observations of shock sensitivity. Generally, regression techniques require satisfying the Central Limit Theorem requirements which demand a higher level of observations. Unfortunately, the data is limited by the number of materials available in the study.

The relationship between the surface roughness characteristics of the RDX materials and their shock sensitivity is investigated. A simple plot (Figure 8) shows that the shock sensitivity of the material is correlated with the level of surface roughness.

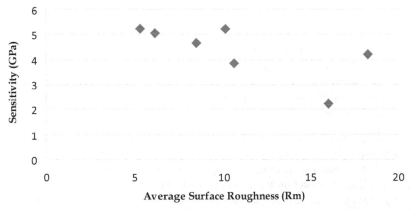

Fig. 8. Plot of sensitivity of material versus average surface roughness (Rm)

However, the plot also demonstrates that there is potential for heteroscedasticity in the data. Although heteroscedasticity does not create a bias in determining the regression coefficients themselves, the statistical significance of those coefficients becomes essentially unknown as the standard errors become incorrectly computed by the basic OLS technique. To test for the presence of heteroscedasticity the Breusch-Pagan / Cook-Weisberg test is employed with the test statistics distributed as χ^2 with the degrees of freedom equal to the number of regressors. The Breusch-Pagan / Cook-Weisberg test for heteroscedasticity indicates $\chi^2(1)=3.71$, which corresponds to a probability $> \chi^2 = 0.0542$. Thus, with a p value of less than 0.05 the test fails to reject the hypothesis of no heteroscedasticty, however with a p value near 0.1, the hypothesis of no heteroscedasticty is rejected. In an effort to be conservative in the analysis a value of p=0.1 is selected. The problem is further amplified by the fact that the sample size is very small (only seven observations of shock sensitivity), which reduces the robustness of the White/Huber estimator, a commonly used method for heteroscedasticity correction. As a result, the HCCM estimator known as HC3 is employed, this estimation technique was discussed by MacKinnon and White (1985), and was later shown to perform better than its alternatives in small samples, see Long and Ervin (2000).

The test shows that the relationship between shock sensitivity and S_m does not exhibit any heteroscedasticity. The Breusch-Pagan / Cook-Weisberg test provides $\chi^2(1)=1.73$, which corresponds to probability $> \chi^2 = 0.189$. The test also finds no heteroscedasticity issues in the relationship between shock sensitivity and $S_{m\mu}$.

The regression analysis in essence plots a best fit line through the data plot, a line that minimizes the sum of squares in the differences between the predicted line Y values and the actual Y values. Table 6 presents the regression output for several specifications. In all of the reported specifications the dependent variable is the level of sensitivity (GPa). The coefficients are reported along with their corresponding t values, included below the coefficient.

Specification I simply examines the impact Rm has on the level of sensitivity of the material. This specification is equivalent to simply plotting the best fit line through the dataset in Figure 1 and is estimated using the HC3 method for the computation of errors. The model can be summarized by the following equation:

$$Sensitivity = 5.998 - 0.153R_m$$

However, the statistical validity of that equation is limited. First, the goodness of fit is low. As measured by the adjusted R-squared, the model explains only about 8% of volatility in the level of sensitivity. Secondly, the coefficient on R_m is statistically significant at only 67%. All the subsequent specifications are estimated without the HC3 technique, as no evidence of heteroscedasticity was found (see the discussion above).

Specification II models Sensitivity as a function of Sm. The level of statistical significance increases substantially. The coefficient on S_m is statistically significant at 99%, suggesting that the level of variation in surface roughness on the surface of the particle has a statistically significant impact on the level of shock sensitivity of the material. Furthermore, the overall goodness of fit of this specification has also improved. The adjusted R-squared suggests that model in Specification II explains over 76% of volatility in sensitivity.

Specification III examines the relationship between $S_{m\mu}$ and sensitivity. The coefficient on $S_{m\mu}$ is statistically significant at 92%. The overall fit of the model is also weaker with the adjusted R-square being at 0.398.

In Specification IV these three measures of surface characteristics are combined into one model. The test for heteroscedasticity failed to reject the hypothesis of homoscedasticity. Thus, the estimation is estimated without the HC3 technique. The overall explanatory power of the model improves substantially with the adjusted R-squared rising to 0.915. The coefficient on R_m is statistically significant at only 67% and the one on $S_{m\mu}$ is statistically significant at 84%, but the coefficient on S_m remains statistically significant at 98%. This specification includes multiple measures of surface characteristics that may in tern be correlated with each other. Although this has already been examined in the correlation table and no meaningful correlation was observed, this is verified with a computation of the variance inflation factor (VIF), see Table 7. The VIF computation confirms what the correlation table suggested, no multicollinearity problems in specification IV.

| | Dependent Variable = Sensitivity | | | | |
| | Specification | | | | |
Ind. Vars.	I	II	III	IV	V
Rm	-0.15			-0.04	
	1.07			1.14	
Sm		-0.29		-0.23	
		4.52		5.08	
Smμ			-0.55	-0.23	
			2.23	1.87	
%HMX					-0.05
					0.41
Constant	6.00	6.02	7.26	7.31	4.47
	5.31	14.34	5.41	14.38	8.39
R-sq	0.48	0.80	0.50	0.96	0.03
Adj. R-sq	0.08	0.76	0.40	0.92	-0.16

Table 6. OLS Regression with Heteroscedasticity correction

Variable	VIF	1/VIF
Rm	1.89	0.530049
Smμ	1.81	0.552178
Sm	1.3	0.771688

Table 7. Computation of Variance Inflation Factor.

Specification IV can be written as a simple equation where each coefficient can be interpreted as a partial derivative:

$$Sensitivity = 7.308 - 0.042R_m - 0.225S_m - 0.233S_{m\mu}$$

The last specification examines the relationship between the level of impurity (%HMX) and sensitivity. The regression analysis demonstrates that there is no statistically significant relationship between these two variables. For further analysis of shock sensitivity and its determinants in RDX materials see Bellitto and Melnik (2010) and Bellitto et al (2010).

4. Conclusion

Atomic force microscopy can be used to obtain a large number of data observations at the nanometric level. These data can statistically be used to investigate and establish quantitative relationships between various variables. This work demonstrates that surface characteristics data obtained from topographical scans can be used in investigating the relationship between the shock sensitivity of the materials at the macroscale and their surface roughness characteristics at the nanoscale. Statistical analysis can be used not only to show that there is a statistical relationship, but with the help of regression techniques, can precisely estimate any such relationships. As demonstrated in this chapter, the surface roughness variation on the surface of the particle has a substantial negative impact on the shock sensitivity in RDX materials. A one unit increase in the average standard deviation (S_m) reduces the shock sensitivity by 0.225 GPa. The statistical analysis can also be used to demonstrate absence of any meaningful relationship. For instance, the results demonstrate that the level of HMX impurity does not impact the shock sensitivity in the studied RDX materials.

5. Acknowledgements

This work was supported by the Office of Naval Research (ONR) and the Naval Surface Warfare Center (NSWC) at Indian Head, MD. The authors express their thanks to Mary Sherlock, Robert Raines, Tina Woodland and Philip Thomas for helpful discussions and for providing the samples.

6. References

Doherty, R..M. & Watts, D.S. (2008). Relationship between RDX Properties and Sensitivity. *Propellants, Explosives, Pyrotechnics*, Vol.33, pp. 4-13
Bellitto, V.J. & Melnik M.I. (2010). Surface defects and their role in the shock sensitivity of cyclotrimethylene-trinitramine. *Applied Surface Science*, Vol.256, pp. 3478-3481
Bellitto, V.J.; Melnik, M.I.; Sorensen, D.N. & Chang, J.C. (2010). Predicting the Shock Sensitivity of Cyclotrimethylene-Trinitramine. *Journal of Thermal Analysis and Calorimetery*, Vol. 102, pp. 557-562
Greene, W. H. (2003). *Econometric Analysis*, 5th edition, Prentice Hall,
Levine, D.M.; Stephan, D. F.; Krehbiel, T.C. & Berenson, M.L. (2011). *Statistics for Managers using Microsoft Excel*, 6th edition, Prentice Hall
Long, J.S. & L.H. Ervin (2000). Using Heteroscedasticity Consistent Standard Errors in the Linear Regression Model. *The American Statistician* Vol. 54, pp. 217-224
MacKinnon, J.G. & White, H. (1985). Some heteroskedasticity consistent covariance matrix estimators with improved finite sample properties. *Journal of Econometrics*, Vol . 29, pp. 53-57

9

AFM Application in III-Nitride Materials and Devices

Z. Chen[1], L.W. Su[2], J.Y. Shi[2], X.L. Wang[2], C.L. Tang[2] and P. Gao[2]

[1]Golden Sand River California Corporation, Palo Alto, California
[2]Lattice Power Corporation, Jiangxi
[1]USA
[2]China

1. Introduction

The nitride family is an exciting material system for optoelectronics industry. Indium nitride (InN), gallium nitride (GaN) and aluminium nitride are all direct bandgap materials, and their energy gaps cover a spectral range from infrared (IR) to deep ultraviolet (UV). This means that by using binary and ternary alloys of these compounds, emission at any visible wavelength should be achievable.

Atomic force microscopy (AFM) is a powerful tool to study the III-nitride surface morphology, crystal growth evolution and devices characteristics. In this chapter, we provide an overview of AFM application in AlInGaN based materials and devices.

Typical surface morphologies of GaN materials characterized by AFM are presented in §1. In additional, three types of threading dislocations, including edge, screw and mixed threading dislocation, are studied by AFM in this section. In §2, V-shape defects and other features in InN, InGaN film and InGaN/GaN multiple quantum wells are summarized. It is not easy to grow high quality AlN and AlGaN films, materials for UV light emitting diode and high electron mobility transistor, which usually have high dislocation density and three dimensional growth mode. Growth condition optimization of high quality, crack-free and smooth Al(Ga)N, with the assistance of AFM, is reviewed in §3. Applications of AFM in GaN based devices are discussed in §4, including the patterned sapphire substrates, as-grown LED surface, backside polished surface, ITO surface investigation.

2. AFM study of GaN

In recent years, the III-nitride-based alloy system has attracted special attention since high-brightness blue, green and white light-emitting diodes (LEDs), and blue laser diodes (LDs) became commercially available.

Good topography and crystal quality of GaN films are some of the key factors to improve devices performance. AFM helps researchers to observe the surface morphology of GaN, to study dislocations in GaN film, and to further understand and optimize the growth condition of the GaN.

2.1 AFM study of GaN surface morphology

Fig. 1(a) is a $10 \times 10\ \mu m^2$ AFM image of typical high-quality, fully coalesced GaN film on sapphire grown by metal organic chemical vapor deposition (MOCVD). Well-defined, uniform and long crystallographic steps could be observed on the surface. The sample topography exhibited atomic terraces with a measured height of 0.3 nm, in close agreement with the 0.26 nm bi-layer spacing of GaN. Fig. 1(b) is $20 \times 20\ \mu m^2$ AFM image of poor-quality, uncoalesced GaN film for comparison. The biggest void (hole) was formed at the coalescence boundaries of the nanoisland growth fronts. The width of the observed hole is up to 2.5 μm and the depth could be several microns because the holes result from the poor coalescent at the beginning of the growth.

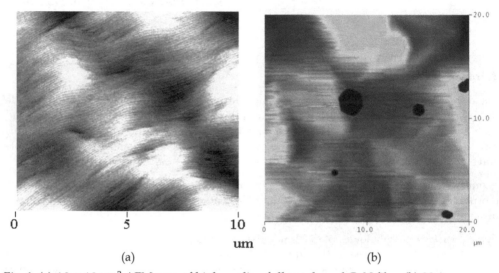

(a)   (b)

Fig. 1. (a) $10 \times 10\ \mu m^2$ AFM scan of high quality, fully coalesced GaN film; (b) $20 \times 20\ \mu m^2$ AFM scan of un- coalesced GaN film.

The intersection of a screw-component dislocation with the film surface creates an atomic step termination that may lead to spiral step procession and hillock formation (Burton et al 1951). At the top of these hillocks AFM reveals the presence of either single screw or clustered screw and mixed-type dislocations, as shown in Fig. 2(a). The presence of clustered defects with the screw component of the Burgers vector is a reason of formation of the growth hillocks.

Spiral growth hillocks, with pits located at the top and some pits located away from the hillock peaks, are shown in Fig. 2(b). Pits at the center of a spiral growth hillock are expected to have a screw component, while dislocation pits away from hillock peaks without terminated steps are of pure edge character.

2.2 Study of dislocation in GaN by AFM

Fig. 3 is a $2 \times 2\ \mu m^2$ AFM image of a GaN film, in which three kinds of pits of different sizes correspond to three different dislocations were observed. Three types of threading

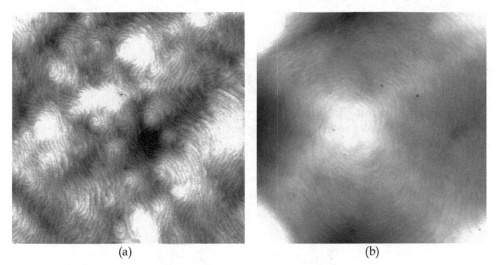

Fig. 2. 5 × 5 μm² AFM scans of the (a) GaN sample with lot of hillocks; (b) GaN sample with single hillock and several pits.

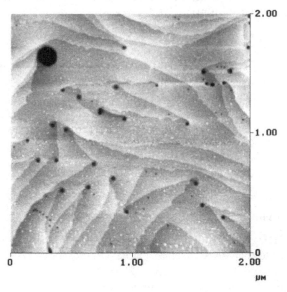

Fig. 3. AFM image of one GaN film showing three types of pits, corresponding to three kinds of dislocations in GaN.

dislocations, edge, screw and mixed types are usually observed in wurtzite GaN epitaxial layers, with the corresponding Burgers vectors, confirmed by TEM (Follstaedt et al. 2003 ; Datta et al. 2004).

$$b_{edge} = 1 / 3 \left[11\bar{2}0 \right]$$

$$b_{mixed} = 1/3\left[11\overline{2}3\right]$$

$$b_{screw} = [0001]$$

According to the thermodynamics of pit formation (Sangwal, 1987), it follows that the potential difference ($\Delta\mu$ of a stable nucleus of a pit depends inversely on the elastic energy (E_{el}) of the dislocation: $\Delta\mu = 2\pi^2\Omega\gamma^2 / E_{el}$ where: E_{el} is elastic energy of dislocation, γ is edge free energy, Ω is molecular volume. The elastic energy value for screw, edge and mixed type of dislocations are:

$$E_{screw} = Gb^2\alpha$$

$$E_{edge} = Gb^2\alpha(\frac{1}{1-v})$$

$$E_{mixed} = Gb^2\alpha(1 - v\cos\theta / (1-v))$$

(where: G is shear modulus, b is Burgers vector, α is geometrical factor, v is Poisson's constant and θ is the angle between screw and edge components of the Burgers vector of mixed dislocations) (Hull et al., 1984).

Large differences in the magnitude of Burgers vectors, especially between edge type and screw/mixed type dislocations, imply that the size of pits should be different depending on the type of dislocation, i.e. the largest pits are formed on screw-, intermediate size pits on mixed- and the smallest ones on edge-type dislocations (Weyher et al., 2004). The densities of the pits in the sample shown in Fig. 3 with median and larger sizes are $7.5 \times 10^8/cm^2$, in agreement with the expected density of screw component dislocations. The majority of dislocations are of pure edge character, related to the smallest pits, with a density of $1 \times 10^9/cm^2$, in agreement with the expected pure edge dislocation density determined by TEM examination.

3. AFM study for In(Ga)N

3.1 V-shape defect in InGaN

Bulk InGaN or InGaN/GaN multiple quantum wells (MQWs) have been used as active layers for near UV, blue, green and white LEDs, laser diodes and solar cells due to the tunable band-gap energy of InGaN, from 0.7 to 3.4 eV through indium composition variation. Therefore it is important to understand the role of microstructural and compositional inhomogeneities in the InGaN layers on the optical emission.

InGaN MQWs structures often have a so-called V-shape defect (Wu et al., 1998; Kim et al., 1998; Northrup et al. 1999; Duxbury et al., 2000; Scholz et al., 2000; Kobayashi et al., 1998) that consists of a threading dislocation terminated by a pit in the shape of an inverted hexagonal pyramid with $[10\overline{1}1]$ sidewalls, as shown in Fig. 4. Pit formation creates 6 equivalent $[10\overline{1}1]$ facets. The depth of the pit is h=1.63a. The angle of the hexagonal inverted pyramid defects is 61°, which could be measured accurately by AFM shown in Fig. 4.

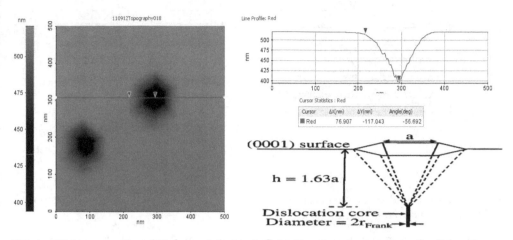

Fig. 4. AFM topography of V-shape defect in InGaN film. A representation of a dislocation terminating in a pit at the (0001) surface is shown.

It is usually accepted that the high defect density in GaN leads to poor optical property and also affects the structural and optical quality of the active layer composed of the InGaN/GaN MQWs. It has been reported that threading dislocations disrupt the InGaN/GaN MQW and initiate the V defect using transmission electron microscopy (TEM) and atomic force microscopy (AFM) (Sharma et al., 2000 ; Lin et al., 2000). Several research groups have reported that there is always a threading dislocation (TD) connected with the bottom of V defect and the cause of V-defect formation is the increased strain energy and the reduced Ga incorporation on the [10$\bar{1}$1] pyramid planes compared with the [0001] plane (Sun et al., 1997). Cho et al. investigate the V defects in InGaN/GaN MQWs by TEM (Fig. 5) and found that the origin of V defects are not only connected to TD, but also generated from the stacking mismatch boundaries (SMBs) induced by stacking faults (SFs) shown in Fig. 6(a) and 6(b) (Cho et al., 2001).

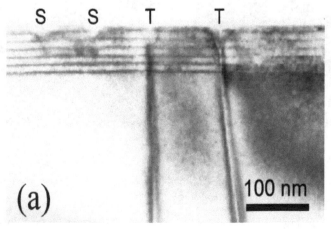

Fig. 5. Cross-sectional bright-field TEM images of the In$_{0.3}$Ga$_{0.7}$N/GaN MQWs.

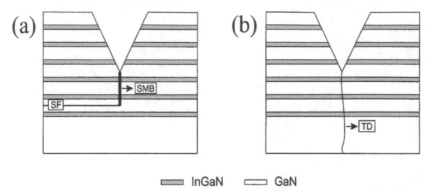

Fig. 6. Schematic models for V-defect formation connected with (a) a threading dislocation and (b) a SMB induced by stacking faults.

The AFM images in Fig. 7 shows the $5 \times 5\ \mu m^2$ scans of the InGaN/GaN superlattice (SL) structure, which is used as the strain release layer in LEDs and LDs. Sample A, with In% composition of 8% is pictured in Fig. 7(a) and sample B, with In% composition of 2% is shown in Fig. 7(b). Remarkable morphological differences are noticeable. A slightly lower density of pits with a wider range of pit diameters is noted for sample A (Fig. 7(a)] when compared to sample B (Fig. 7(b)). In addition, the typical inclusions are shown for sample A as bright, irregular white features. The V-pits density in Fig. 7(a) is around $3 \times 10^8/cm^2$, with the average diameter of around 120 nm, and inclusion heights up to 120 Å. The defect density in sample B (less Indium) is around $4 \times 10^8/cm^2$ with diameters in the 50 nm range.

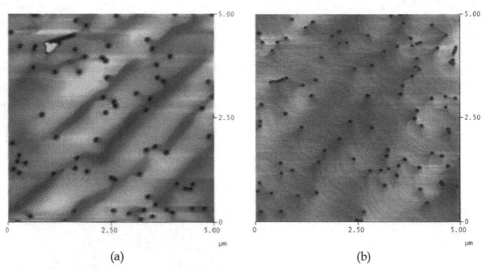

Fig. 7. $5 \times 5\ \mu m^2$ AFM scans of the InGaN/GaN superlattice (SL) structure with (a) indium composition of 7% (b) indium composition of 2%.

The AFM images in Fig. 8 show the $5 \times 5\ \mu m^2$ scans of a InGaN/GaN MQW structure. As previously described in Fig. 4, a V-pit usually connected to a TD and the diameter of the TD

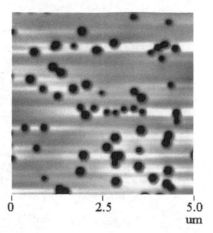

Fig. 8. 5 × 5 µm² AFM scans of a InGaN/GaN MQW structure. Two kinds of V-defects with different sizes were observed.

should depend on the magnitude of the Burgers vector of dislocations. The bigger TD diameter results in a bigger pit, thus, the bigger V-pits on the InGaN/GaN MQWs are rooted to screw or mixed TD, while the smaller V-pits connected to edge TDs. The diameter of the bigger V-pits in Fig. 8 is about 300 nm, with the density around $1.2 \times 10^8/cm^2$, corresponding to the density of the screw TD, while diameter and density of the smaller V-pits in Fig. 8 is about 180 nm and $1.0 \times 10^8/cm^2$, corresponding to the density of the edge TD. The size of the V-pits is bigger than InGaN/GaN SL structure shown in the previous Fig. 7(a) because V-defects grow bigger and bigger with a thicker InGaN layer and a higher Indium composition. The density of the V-pits usually is lower than the TD density in GaN underlayer because not every TD develops to a V-pit. In additional, only one kind of V-pit with the same size was observed for some InGaN/GaN sample, because the V-pits connected to screw TDs is easier to be opened.

3.2 InN growth condition optimization

InN has a band gap value of 0.67–0.8 eV, which potentially extends the spectral range covered by group-III nitrides to the near-infrared. In addition, InN has a very small electron effective mass and a high electron drift velocity (O'Leary et al., 1998). Recent theoretical calculations predict an ultimate room temperature electron mobility for InN to reach 14,000 cm²/Vs. (Polyakov et al., 2006). However, it is very difficult to grow InN materials due to the thermal instability of InN and the large lattice mismatches between InN and substrates. A high V/III flux ratio and a low growth temperature are usually required to suppress InN decomposition, which often results in unsatisfactory crystal quality and undesired three-dimensional (3D) surfaces roughness. Several technologies have been applied to improve InN film quality and surface morphology. However, AFM has been the most common tools for surface morphology evaluation and improvement.

Fig. 9 is an AFM scan showing the surface morphology of a typical InN film grown by MOCVD on GaN. A low density of large islands with widths up to about 1 µm and heights of 300–400 nm can be observed, which might be indium droplets (N"orenberg et al., 2002).

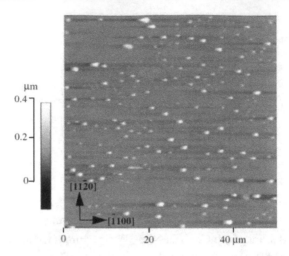

Fig. 9. AFM micrographs of typical InN layer.

InN is usually grown by MBE due to its low growth temperature. Chad et al. grew InN films in two different growth regimes and characterized the surface morphology by AFM, as shown in Fig. 10(a) (the N-rich regime) and Fig. 10(b) (the In-droplet regime). Clear growth steps were observed for InN grown in N-rich regime and Indium droplets were observed when it was grown in the In-droplet regime (Gallinat et al., 2006).

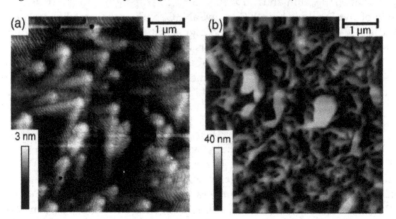

Fig. 10. AFM micrographs of (a) a 1.5- μm -thick InN layer grown in the In-droplet regime and (b) a 1- μm -thick InN layer grown in the N-rich regime.

4. AFM study of Al(Ga)N

4.1 AlN growth condition optimization

Due to its potential for many applications, the growth and characterization of AlN has been a subject of much interest. Great progresses on AlN material growth and device fabrication have been attained. The first electroluminescent emission at wavelength of 200 nm from a p-i-n AlN homojunction LED operating at 210 nm was realized (Taniyasu et al., 2006). The first AlN metal-semiconductor-metal photodetector with a peak responsivity at 200 nm was. obtained (Li et al., 2003). Significant improvement of the AlN crystal quality using Lateral Epitaxial Overgrowth (LEO) of AlN on sapphire substrates was also demonstrated (Chen et al., 2006). This resulted in 214 nm stimulated emission, the shortest wavelength stimulated emission reported in semiconductor materials (Shatalov et al., 2006).

AlN is widely used as the buffer layer for Ultraviolet (UV) Light Emitting Diodes (LED) and High Electron Mobility Transistors (HEMT) grown on SiC substrates currently due to its small lattice constant, wide bandgap and high thermal conductivity (Chen et al. 2009a; Chen et al. 2009b). It is therefore necessary to develop an uncomplicated method for growing high quality, thick and crack-free AlN on SiC substrates. AFM images of the AlN grown in 3D and 2D modes are shown in Figs. 11(a) and 11(b), respectively. Islands are present in Fig. 11(a), showing that this material has a 3D growth mode. Steps with a height difference of one (0001) AlN monolayer can be seen in Fig. 11(b), showing that the growth mode of this material is 2D. The RMS surface roughness of the 3D AlN over $5 \times 5 \; \mu m^2$ area is 0.517 nm, while that of the 2D AlN is 0.151 nm.

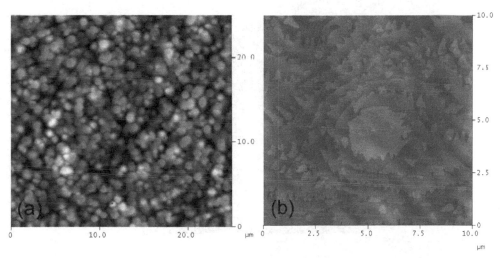

Fig. 11. AFM images of AlN grown at different growth modes: a) three-dimensional, b) two-dimensional.

High quality AlN films were grown by switching between the established 2D and 3D AlN growth modes, a method we call modulation growth (MG) (Chen et al. 2008;). The structure of a MG AlN sample is shown in Fig.12. First, a 300 nm 3D AlN layer was grown on the SiC substrate. Then a 2D 200 nm AlN layer was grown. Subsequently, this 3D-2D period was repeated twice. The total thickness of the resulting film is 1.5 μm. The surface of the AlN grown by 3D-2D modulation growth is very smooth. Fig. 13 shows a $5 \times 5 \; \mu m^2$ AFM scan of the surface with an RMS surface roughness of 0.132 nm. Well defined steps and terraces indicate a step-flow growth mode, and can be observed in all parts of the wafer. The height difference between terraces corresponds to one monolayer of (0001) AlN. No step terminations were observed over the scanned area indicating a low density of Threading

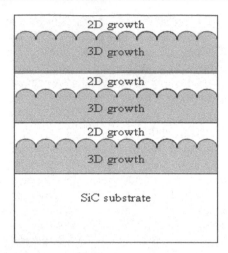

Fig. 12. Schematic of AlN grown with 3D-2D modulation growth method.

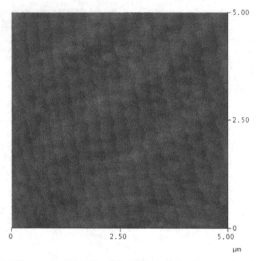

Fig. 13. AFM image of AlN grown with the 3D-2D modulation method.

Dislocation (TD) with screw character, which was confirmed by on-axis and off-axis X-ray rocking curves. The Full Widths at Half Maximum (FWHM) of the (002) and (102) peaks were 86 and 363 arc sec, respectively. The off-axis (105) and (201) peaks were also measured and have FWHM of 225 and 406 arc sec, respectively. The narrow symmetric and asymmetric FWHM of the X-ray data for AlN suggest a low threading dislocation (TD) density, with a small proportion of TDs with screw components.

An atomically flat surface can be obtained when AlN growth is performed at temperatures higher than 1300 °C (Imura et al., 2007). However, the dislocation density is still as high as $1 \times 10 \ cm^{-10}$. Therefore, the crystalline quality of AlN requires further improvement, particularly when it is grown on sapphire substrates.

Epitaxial lateral overgrowth (ELOG) and similar techniques have proved to be an effective method to reduce the TD in GaN (Detchprohm et al., 2001; Weimann et al., 1998) and improve the devices performance. However, ELOG is difficult to undertake with AlN, and even with AlGaN, due to the high sticking coefficient of Al adatom, thus causing a low lateral growth. For the first time, Chen et al. reported the pulsed lateral epitaxial overgrowth (PLOG) of AlN films over shallow grooved sapphire substrates (Chen et al., 2006). The PLOG approach at temperatures around 1150 °C enhances the adatom migration thereby significantly increasing the lateral growth rates. This enables a full coalescence in wing regions as wide as 4 to 10 μm.

In the ElOG process for AlN, first a 0.3-μm-thick AlN layer was grown on the sapphire substrates using a migration-enhanced low pressure MOCVD process. Standard photolithography was then used to form 2 μm wide-masked stripes, with 4-10 μm wide openings oriented along the AlN $< 1\bar{1}00 >$ directions. Reactive ion etching was used to remove the AlN and sapphire in the openings area to form 4-10μm-wide and 0.7-μm-deep trenches. This grooved sapphire/AlN sample served as the template for the subsequent pulsed lateral epitaxial overgrowth. TMA and NH₃ were used as the precursors, and growth was carried out at 1150 °C. For the PLOG process, the NH₃ supplied to the reactor was pulsed (pulse durations 6 to 12 seconds) while the TMA flow was kept constant during the growth.

Fig. 14 shows a cross-sectional SEM image of a fully coalesced PLOG-AlN film with a 4- μm -wide trench (4- μm -wide support mesa). The SEM image clearly shows lateral growth in the $< 11\bar{2}0 >$direction. The sidewall and the top surface are very smooth. A complete triangular void forms upon the trench after the coalescence of the stripes when thickness of the film reaches 6 μm . This indicates the lateral to vertical growth rate ratio is 1:3.

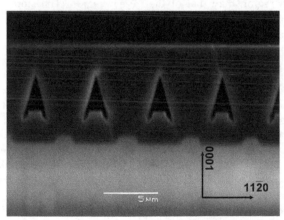

Fig. 14. Cross sectional SEM image of AlN film grown on grooved template by PLOG.

After coalescence, the AlN surface became flat over the whole wafer. A $10 \times 10\ \mu m^2$ atomic force microscopy (AFM) scan shows an RMS surface roughness of 0.5 nm. Fig. 15 shows a $6 \times 6\ \mu m^2$ AFM scan image covering one mask period of the PLOG-AlN surface including both the region directly grown on top of the support mesa and the region laterally grown over the trench. As seen from the well-defined steps and terraces, a two dimensional step-flow growth is dominant over the entire wafer under these PLOG growth conditions.

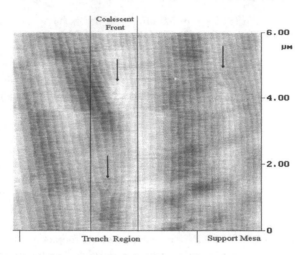

Fig. 15. AFM image acquired from the PLOG-AlN trench and support mesa regions. Arrows point to intersections of dislocations at the surface.

Height differences of 0.27 nm between terraces correspond to one monolayer of (0001) AlN (c/2=0.25nm). The overgrown region has long parallel atomic steps without step terminations, indicating a reduced TD density. At the coalescence point, two threading dislocations marked by arrows are seen in Fig. 15 as step terminations in the AFM image. Another TD is observed in the support area in the same image, implying higher TD density in the support mesa region than that of the trench region. The step termination density, corresponding with either pure screw or mixed screw-edge character, measured by AFM, is $8.3 \times 10^6 \text{cm}^{-2}$ for the scan area shown in Fig. 15.

4.2 AlGaN material growth

The growth of AlxGa1-xN can be carried out on sapphire substrates with a thin low temperature (LT) AlN nucleation layer (Wickenden et al., 1998; Wang et al., 2007 ; Koide et al., 1988) or thick high temperature (HT) AlN template (Sun et al., 2004 ; Mayes et al., 2004 ; Fischer et al., 2004). Grandusky et al. studied the effect of LT-AlN nucleation layer growth conditions for the growth of high quality AlxGa1-xN layers on sapphire (Grandusky et al., 2007). The conditions of the LT AlN had a dramatic effect on the morphology and crystalline quality of the overgrown AlxGa1-xN layers. The effect of growth temperature and thickness of the LT AlN nucleation layers on the overgrown Al0.5Ga0.5N is shown in Fig. 16. Fig. 16(a) shows a smooth surface with a RMS roughness of 1 nm over 10 × 10 μm, whereas Figs. 16(b) and (c) show rough surfaces with 3D crystallites clearly evident. As demonstrated in Fig. 16, extreme changes in the surface morphology are seen with further increase in temperature of the nucleation layer (NL). For a lower growth temperature of 525 °C for the AlN NL, little change was seen in the surface morphology. The best surface morphology was the sample with a thickness of 15 nm as can be seen from Figs. 16(d)–(f). Large AlGaN surface crystallites were observed when the nucleation layer was too thin, as shown in Fig. 16(d). When the layer was thicker than 15 nm, slight roughening of the surface could be seen in Fig. 16(f), as well as a broadening of the (0002) rocking curve.

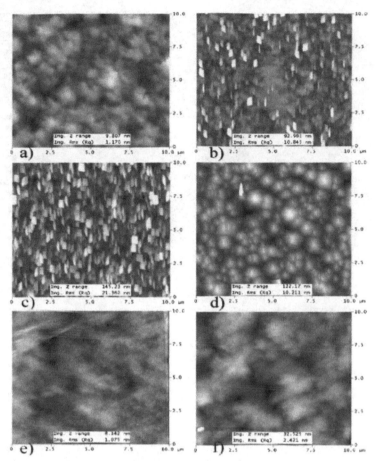

Fig. 16. AFM scans of 500 nm thick Al0.5Ga0.5N grown on a 15 nm AlN nucleation layer grown at (a) 625 °C, (b) 650 °C, and (c) 675 °C and on a low temperature 525 °C AlN NL with thicknesses of (d) 10 nm, (e) 15 nm, and (f) 30 .

Subsequently, it was observed by AFM that the direct growth of an 1μm layer of AlGaN on AlN homoepitaxial layers always results in a roughened morphology. To facilitate heteroepitaxial strain relaxation while preserving structural quality. Z. Ren et al. adopted a design of step-graded layers consisting of three superlattices (SLs 1, 2, and 3) with average Al compositions of 0.90, 0.73, and 0.57. Each SL was composed of ten periods of $Al_xGa_{1-x}N$ (150 Å)/ $Al_yGa_{1-y}N$ (150 Å) (x/y=1.0/ 0.8, 0.8/ 0.65, and 0.65/ 0.50) (Ren et al., 2007). Surface morphology after the growth of SL 1 (Figs. 17(a) and (d)) and SL 2 (Figs. 17(b) and (e)) indicates that pseudomorphic growth persists with an atomically smooth surface under a step-flow growth mode. During the growth of SL 3, surface morphology underwent a fundamental change with the appearance of large 1–2 μm plateaus or platelets separated by deep trenches or clifflike edges with a height of ~100 nm (Fig. 17(c)) even though step flow was still maintained locally (Fig. 17(f)) (Ren et al. 2007).

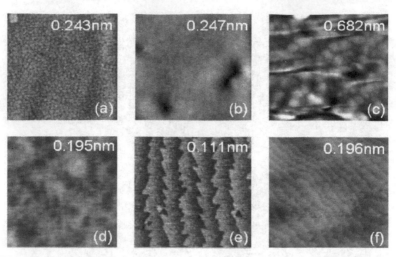

Fig. 17. AFM images of AlGaN superlattices on bulk AlN. The scan area of (a), (b), and (c) is 10×10 μm^2, and (d), (e), and (f) is 1×1 μm^2. (a) and (d), (b) and (e), and (e) and (f) are taken after the growth of superlattices 1, 2, and 3, respectively. Root-mean-square (rms) roughness from each scan is labeled.

5. Investigation of III-nitride devices by AFM

The structure of a lateral blue or green InGaN LED is shown in Fig. 18. To observe the surface of each layer and know what happened after each device processing step, researchers use SEM and AFM to check the surface morphology of the patterned sapphire substrates, as-grown LED wafers, etched n-mesa, ITO and p/n metal.

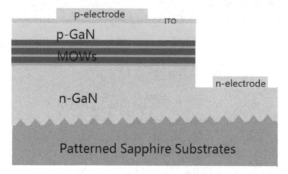

Fig. 18. Structure of a typical lateral GaN based light emitting diode.

5.1 Patterned sapphire substrates

Most of the current commercial LEDs are grown on patterned sapphire substrates (PSS), not on flat sapphire substrates, because LEDs on PSS have demonstrated an enhanced light output power and external quantum efficiency (EQE) compared to conventional LEDs grown on planer sapphire substrates.

LED grown on PSS showed a higher internal quantum efficiency due to a dislocation density reduction by epitaxial lateral overgrowth technology (Tadatomo et al., 2001; Yamada et al., 2002 ; Hsu et al., 2004). Besides the elimination of threading dislocations due to the lateral growth of GaN on top of PSS, researchers believe that PSS improve the light extraction. It is well known that the large difference of refractive index between semiconductor and air leads to trap of large percentage of light emitting from LED (Lee et al., 2005). Thus, it is important to design and characterize the geometry of the PSS. AFM is the only available tool to characterize the bump shape accurately so far.

AFM images of PSS with different specs are shown in Figs. 19(a) and (b). The bump shapes, depths and widths could be characterized accurately. The average bump depth, pitch and gap of the patterns seen in Fig. 19(a) are 1.1 µm, 3 µm and 0.2 µm, respectively. While the average bump depth, pitch and gap of the patterns seen in Fig. 19(b) are 0.3 µm, 6 µm and 3 µm, respectively.

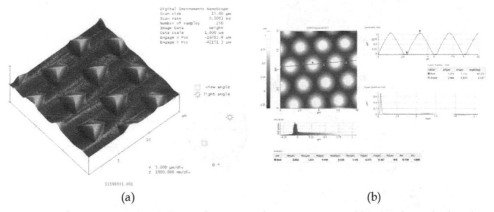

(a)   (b)

Fig. 19. (a) One patterned sapphire substrate with a cross sectional line to show the height, width and shape of the bumps. (b) 3-dimensional AFM image of another PSS with triangular pyramid cone shaped bumps.

5.2 As grown LED surface and leakage characteristics

Fig. 20 (a) is the surface morphology of an LED grown on a PSS. There is a p++ layer on the top of the LED surface. The bumps on the surface are caused by heavily doped Mg on the surface, usually rooted from a screw type dislocation. In the manufacturing of LEDs it is important to verify that the surface is free of pits, which implies the p-layer has coalesced well and sealed all the V-pits from the underneath layer of InGaN. Fig. 20(b) is an example of a LED with pits on the surface. LEDs with un-coalesced surfaces with pits usually have high leakage currents.

Screw type related pits on the GaN surface have been confirmed to be the source of reverse leakage in GaN films (Law et al., 2010). Fig. 21(a) shows an AFM topograph of a GaN sample grown by molecular beam epitaxy (MBE), in which the screw type TD related pits could be observed. Fig. 21(b)–(d) show conductive atomic force microscopy (CAFM) images obtained at dc biases of -14, -18, and -22 V of the area in Fig. 21(a). Fig. 21(b) shows several small, dark features that correspond to localized reverse-bias leakage paths observable

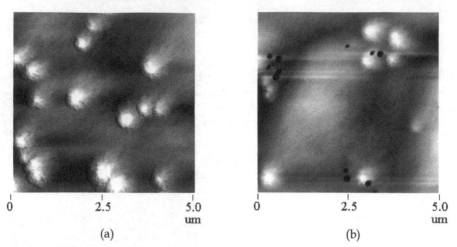

<table>
<tr><td>0</td><td>2.5</td><td>5.0</td><td>0</td><td>2.5</td><td>5.0</td></tr>
<tr><td></td><td></td><td>um</td><td></td><td></td><td>um</td></tr>
<tr><td></td><td>(a)</td><td></td><td></td><td>(b)</td><td></td></tr>
</table>

Fig. 20. 5 × 5 μm² AFM scans of the (a) fully coalesced LED sample and (b) partially coalesced LED sample.

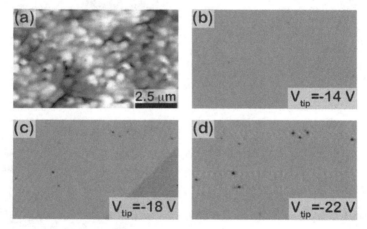

Fig. 21. (a) AFM topograph and (b)–(d) CAFM images obtained at tip dc bias voltages of -14, -18, and -22 V for MBE GaN sample.

at -14 V bias. In Fig. 21(c), the reverse-bias voltage magnitude was further increased and the density of observed conductive paths increased as well. This trend of increasing conductive path density as a function of increasing reverse-bias voltage magnitude continues in Fig. 21(d). According to these results, J. J. M. Law et al. suggested that changes in surface defects surrounding or impurities along screw-component threading dislocations are responsible for their conductive nature (Law et al., 2010).

5.3 Other device processing characterized by AFM

Chemical-mechanical polishing (CMP) is a polishing technique used to thin down the substrates for GaN devices, including Si, SiC, sapphire, AlN and GaN free-standing substrates, because it may produce high quality surface at low cost and fast material-removal rates. Fig. 22 is a typical 25× 25 μm² AFM image of a sapphire substrates polished by CMP.

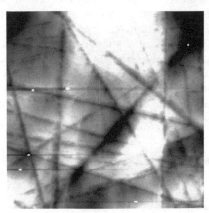

Fig. 22. 25 × 25 μm² AFM scans of the surface topography of a polished sapphire substrate.

Semi-transparent Ni/Au on Mg doped GaN was used as the p-contact material in the earlier LED devices. However, the transmittance of such semi-transparent Ni/Au contact is only around 60 - 75%. Although we could increase the transmittance by reducing Ni/Au metal layer thickness, the contact reliability could become an issue when the contact layer thickness becomes too small. Transparent indium in oxide (ITO) was popularized as the p-contact material because its high electrical conductivity and transparency to visible light. ITO could be deposited by electron beam evaporation or a magnetron sputtering method. It was found that the LED using sputtering ITO has 2-3% higher Light output (Lop) than that of E-beam ITO. Surface morphology and particle size of the ITO film deposited by two methods were measured by AFM as shown in Fig. 23. AFM analysis helped researchers to better understand the quality difference between the two ITO films. Sputtering ITO was shown to have smaller particle and better uniformity.

(a)   (b)

Fig. 23. 5× 5 μm² AFM scans of the ITO deposited by (a) electron beam evaporation and (b) magnetron sputtering.

6. Conclusion

In summary, the application of AFM in GaN, In(Ga)N and Al(Ga)N materials research and device fabrication have been reviewed. In regard to the GaN materials, the threading dislocations, including edge, screw and mixed types dislocations, as well as surface features are investigated by AFM. The study of V-shaped defects and topography in InGaN, InN films and InGaN/GaN multiple quantum wells by AFM has been reviewed. For AlN and AlGaN materials, how to utilize AFM to characterize the films and optimize the growth condition are demonstrated. Results also show that AFM is a powerful tool for device characterization and can shed light on device processing optimization.

7. References

Burton W. K., Cabrera N., & Frank F. C., (1951) *Philos. Trans. R. Soc. London, Ser. A* Vol. 243, pp.299

Chen, Z., Fareed, R.S.Q., Gaevski, M., Adivarahan, V., Yang, J.W., Mei, J., Ponce, F.A., Khan, M.A. (2006), *Appl. Phys Lett.*, Vol.89, pp. 081905.

Chen Z., Newman S., Brown D., Chung R., Keller S., Mishra U. K., Denbaars S. P. and Nakamura S., (2008). *Appl. Phys. Lett.*, Vol.93, pp.191906.

Chen Z., Pei Y., Newman S., Brown D., Chung R., Keller S., Denbaars S. P., Nakamura S. & Mishra U. K., (2009). *Appl. Phys. Lett.*, Vol.94, pp.171117.

Chen Z., Pei Y., Newman S., Brown D., Chung R., Keller S., Mishra U. K., Denbaars S. P., & Nakamura S. (2009). *Appl. Phys. Lett.*,Vol. 94, pp.112108.

Cho H. K., Lee J. Y., Yang G. M. & Kim C. S., (2001) *Appl. Phys. Lett.*, Vol.79, pp.215.

Datta R., Kappers M.J., Barnard J.S., Humphreys C.J., (2004). *Appl. Phys. Lett.* Vol. 85, pp. 3411-3413.

Detchprohm T., Sano S., Mochizuki S., Kamiyama S., Amano H., & Akasaki I. (2001). *Phys. Status Solidi A, Vol.*188, pp.799.

Duxbury N., Bangert U., Dawson P., Thrush E. J., Van der Stricht W., Jacobs K., & Moerman I., (2000). *Appl. Phys. Lett.* Vol.76, pp.1600.

Fischer A.J., Allerman A.A., Crawford M.H., Bogart K.H.A., Lee S.R., Kaplar R.J., Chow W.W., Kurtz S.R., Fullmer K.W., & Figiel J.J. (2004). *Appl. Phys. Lett.*, Vol.84 , pp.3394.

Follstaedt D.M., Missert N.A., Koleske D.D., Mitchell C.C., Cross K.C., (2003) *Appl. Phys. Lett.* Vol. 83, pp. 4797-4799.

Gallinat C. S., Koblmüller G., Brown J. S., Bernardis S., Speck J. S., Chern G. D., Readinger E. D. Shen H., & Wraback M., (2006). *Appl. Phys. Lett.* 89, pp.032109.

Grandusky J. R., Jamil M., Jindal V., Tripathi N., & Shahedipour-Sandvik F. (2007). *J. Vac. Sci. Technol.*, Vol.A 25, pp.441.

Hsu Y.P., Chang S.J., Su Y.K., Sheu J.K., Lee C.T., Wen T.C., Wu L.W., Kuo C.H., Chang C.S., Shei S.C., (2004). *J. Cryst. Growth* Vol.261 pp. 466–470.

Hull D. & Bacon D.J., (1984). *Introduction to dislocations*, Pergamon Press, Oxford, pp 71-90.

Imura M., Nakano K., Fujimoto, N., Okada N., Balakrishnan, K., Iwaya M., Kamiyama S., Amano H., Akasaki I., Noro T., Takagi T., Noro T., Takagi T. & Bandoh A. (2007) *Jpn. J. Appl. Phys.*, Vol. 46, pp.1458.

Kim I.-H., Park H.-S., Park Y.-J., & Kim T., (1998). *Appl. Phys. Lett.* Vol. 73, pp.1634-1636.

Kobayashi J. T., Kobayashi N. P., & Dapkus P. D., (1998). Meeting of The American Physical Society, Los Angeles, pp.16–20.

Koide Y., Itoh N., Itoh, K., Sawaki N., & Akasaki I. (1988). *Jpn. J. Appl. Phys.*, Vol.27, pp.1156 .

Law J. J. M., Yu E. T., Koblmüller G., Wu F., & Speck J. S., (2010). *Appl. Phys. Lett.* 96, 102111

LeeY.J., Hsu T.C., Kuo H.C., Wang S.C., Yang Y.L., Yen S.N., Chu Y.T., Shen Y.J., Hsieh M.H., Jou M.J., Lee B.J., (2005) *Materials Science and Engineering B* Vol.122 pp.184–187

Li J., Nam K.B., Nakarmi M.L., Lin J.Y., Jiang H.X., Carrier P., & Wei S.H. (2003). *Appl. Phys. Lett.*, Vol.83, pp.5163.

Lin Y. S., Ma K. J., Hsu C., Feng S. W., Cheng Y. C., Liao C. C., Yang C. C., Chou C. C., Lee C. M., & Chyi J. I., (2000). *Appl. Phys. Lett.* Vol.77, 2988.

Mayes K., Yasan A., McClintock R., Shiell D., Darvish S.R., Kung P., & Razeghi M. (2004). *Appl. Phys. Lett.*, Vol.84, 1046.

Northrup J. E., Romano L. T., & Neugebauer J., (1999). *Appl. Phys. Lett.* Vol.74, pp.2319.

O'Leary S. K., Foutz B. E., Shur M. S., Bhapkar U. V., L Eastman. F., J., (1998). *Appl. Phys.* Vol.83, pp.826.

Orenberg N. C., Martin M. G., Oliver R. A., Castell M. R., & Briggs G. A. D., (2002). *J. Phys. D: Appl. Phys.* pp.35 615.

Polyakov V. M. & Schwierz F., (2006). *Appl. Phys. Lett.* Vol.88, pp.032101.

Ren Z., Sun Q., Kwon S.-Y., Han J., Davitt K., Song Y. K., Nurmikko A. V., Cho H.-K., Liu W., Smart,J. A. & Schowalter L. J., (2007). *Appl. Phys. Lett.*, Vol.91, pp.051116.

Sangwal K., (1987). *Etching of Crystals*, North-Holland, Amsterdam , pp. 87-160.

Scholz F., Off J., Fehrenbacher E., Gfrorer O., & Brockt B., (2000). *Phys. Status Solidi A* Vol.180, 315.

Sharma N., P Thomas., Tricker D., & Humphreys C., (2000). *Appl. Phys. Lett. Vol.77*, 1274.

Shatalov, M., Chen, Z., Gaevski, M., Adivarahan, V., Yang, J., & Khan, A., (2006). *International Workshop on Nitride Semiconductors*, Kyoto, Japan., pp.101.

Sun C. J., Anwar M. Z., Chen Q., Yang J. W., Khan M. A., Shur M. S., Bykhovski A. D., Liliental-Weber Z., Kisielowski C., Smith M., Lin J. Y., & Jiang H. X., (1997). *Appl. Phys. Lett.* 70, 2978.

Sun, W.H., Zhang, J.P., Adivarahan, V., Chitnis, A., Shatalov, M., Wu, S., Mandavilli, V., Yang, J.W., & Khan, M.A. (2004). *Appl. Phys. Lett.*, Vol.85, pp.531.

Tadatomo K., H Okagawa., Ohuchi Y., Tsunekawa T., Imada Y., Kato M., Taguchi T., (2001) *Jpn. J. Appl. Phys.* Vol.40 pp.L583–L585.

Takeuchi Y., T., Amano H., Akasaki I., Yamada N., Kaneko Y., & Wang S. Y., *Appl. Phys. Lett.* (1998). Vol.72, pp.710

Taniyasu Y., Kasu M., & Makimot, T. (2006). *Nature*, Vol.441, pp.325.

Wang X.L., Zhao D.G., Jahn U., Ploog K., Jiang D.S., Yang H. & Liang J.W. (2007). *J. Phys. D: Appl. Phys.*, Vol.40, pp.1113.

Weimann N.G., Eastman L.F., Doppalapudi D., Ng H.M., & Moustakas T.D. *J. Appl. Phys.*, Vol.83, 3656. (1998).

Weyher J.L., Macht L., KamleG. r, Borysiuk J., Grzegory I., (2003). *Phys. Stat. Sol. (c)*, Vol..3 pp.21.

Wickenden, D.K., Bargeron, C.B., Bryden, W.A., Miragliotta, J., & Kistenmacher, T. J. (1994). *Appl. Phys. Lett.*, Vol.65, pp.2024.

Wu X. H., Elsaas C. R., Abare A., Mack M., Keller S., Petroff P. M., DenBaars S. P., Speck J. S., & Rosner S. J., (1998). *Appl. Phys. Lett.* Vol.72, pp.692-694.

Yamada M., Mitani T., Narukawa Y., Shioji S., Niki I., Sonobe S., Deguchi K., Sano M., Mukai T., (2002). *Jpn. J. Appl. Phys.* Vol.41 pp.L1431–L1433.

10

Atomic Force Microscopy to Characterize the Healing Potential of Asphaltic Materials

Prabir Kumar Das, Denis Jelagin, Björn Birgisson and Niki Kringos
Railway and Highway Engineering, KTH Royal Institute of Technology
Sweden

1. Introduction

Worldwide, asphalt concrete is the most commonly used material for the top layer of pavements. The asphalt mixture's ability to provide the necessary stiffness and strength via its strong aggregate skeleton, while at the same time offering a damping and self-restoring ability via its visco-elastic bituminous binder, makes it a uniquely qualified material for increased driving comfort and flexible maintenance and repair actions. Unfortunately, bitumen supply is diminishing as crude sources are depleted and more asphalt refineries install cokers to convert heavy crude components into fuels. It is therefore becoming of imminent urgency to optimize the lifetime of the virgin bitumen from the remaining available crude sources. With 90% of the total European road network having an asphalt surface or incorporating recycled asphalt mixture in one of its base layers, the annual production of asphalt mixtures in Europe is well over 300 million tonnes. It is therefore fair to state that asphalt mixtures play a significant role in the economic viability and international position of the European pavement industry.

The intrinsic self-restoring ability of some bitumen, often referred to as its 'healing potential', could thereby serve as an excellent characteristic that could be capitalized upon. To date, however, there is still very little fundamental insight into what causes some bitumen to be better 'healers' than others. Even less is known about the resulting impact of this healing potential on the overall lifetime of the pavement. Healing potential is therefore very rarely included into pavement lifetime predictions or brought into the planning of maintenance operations, which is a missed opportunity. This will not change until a better understanding is created about the fundamental healing processes, which would allow for tailoring of bitumen during the manufacturing process and could potentially have a significant impact on an increased pavement service lifetime.

Current CE (European Conformity) specifications for bituminous binders do not contribute to advancing the understanding of the healing properties of the bitumen and even in more academic context; researchers often limit themselves to the performance of fatigue tests with and without rest periods from which an overall measure of the stiffness of the sample is calculated. The fatigued samples with rest periods may show a slower decrease of stiffness than the samples that were continuously fatigued. This ratio is then directly used as a quantification of the healing propensity of the bitumen during the rest periods. Yet it could be argued that part of this ratio can be contributed to the visco-elastic unloading behavior of the material and says very little about the chemo-mechanical healing propensity, nor helps in understanding of the controlling parameters.

Bitumen is a complex mixture of molecules of different size and polarity, for which microstructural knowledge is still rather incomplete. As physical properties of bitumen are largely dependent on this microstructure, the prediction of the performance of asphalt pavements is also directly related to this. A detailed knowledge of microstructure is needed to understand the physico-chemistry of bitumen, which can serve as the direct link between the molecular structure and the rheological behaviour (Lesueur et al., 1996; Loeber et al., 1998). Optical microscopy techniques have been employed for more than three centuries to study the microstructure of materials (e.g. Baker, 1742). Researchers (Lu & Redelius 2005; Hesp et al., 2007) used optical microscopy in asphalt field to have a better understanding and visualization of bitumen microstructures. However, because of the opacity and adhesive properties of bitumen, optical microscopy has not received much attention from the asphalt industry. To overcome some of the limitations of optical microscopy, researcher in the asphalt field have chosen to use scanning probe microscopy such as the atomic force microscopy (AFM). AFM is capable of measuring topographic features at atomic and molecular resolutions as compared to the resolution limit of optical microscopy of about 200nm. Moreover, the AFM has the advantage of imaging almost any type of surface which opens the window for investigating microstructures of different polymers and wax modified bitumen.

2. Asphalt under the AFM – historic overview

In an effort to enhance the understanding and characterization of bitumen microstructure, atomic force microscopy (AFM) techniques have been employed to bituminous materials over the last 15 years. Typical image obtained from AFM scanning is depicted in Fig. 1, where one can easily observe the existing of microstructures in the bitumen matrix.

Loeber et al. (1996, 1998) was the first research group who published a inclusive investigation of bitumen using AFM. In these studies, bitumen samples were prepared by a heat-casting method to preserve the solid-state morphology. They revealed ripple microstructures with several micrometers in diameter and tens of nanometers in height. The authors named those rippled yellow and black strips microstructures "bumble bees" and attributed them to the asphaltenes (Redelius, 2009). Other shapes and textures including networks and spherical clusters were also observed in the study.

Based on Loeber's findings, Pauli et al. (2001) wanted to find the correlation between the "bumble bee" shaped microstructures and the amount of asphaltenes in the bitumen. In this study, they investigated solvent-cast films of bitumen under AFM using both friction-force and tapping mode. From these, they found the same Loeber's bee-shaped microstructures. To confirm that the bee-shaped structures were asphaltenes, they doped bitumen with asphaltenes and observed an increase in the density of bee-shaped microstructures. Their group also observed that the microstructure disappeared after repetitive scanning over the same area (Pauli & Grimes, 2003) and assumed the reason for this phenomenon was due to the red laser light of the AFM which produces heat and softening of the surface.

A research group from Vienna University of Technology (Jäger et al., 2004), investigated five different types of bitumen and also reported randomly distributed bee-shaped structures, which, just like the previous researchers they related to the asphaltenes in the bitumen.

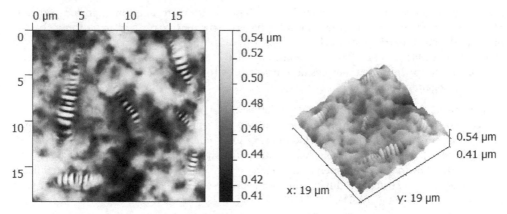

Fig. 1. Topographic 2D (left) and 3D (right) AFM image of bitumen indicating evidence of microstructures

Based on the obtained AFM images from non-contact mode and pulsed-force mode, four different material phases (i.e., hard-bee, soft-bee, hard-matrix and soft-matrix phase) were identified. However, they could not distinguish between soft-bee and soft-matrix phase as the relative stiffness for the two phases were more or less the same. After investigating the bee-shaped structures in the 5μm scale, the authors reported that the distance between the higher parts of the bees was approximately 550nm and appeared to be independent of the source of bitumen.

After this, Masson et al. (2006) conducted extensive AFM studies on the microstructure phases in bitumen by using 12 different types of Strategic Highway Research Program (SHRP) bitumen. In this study, bitumen samples were prepared by heat casting films onto glass plates. By using AFM phase detection microscopy (PDM), the detected four phases were defined as catanaphase (bee-shaped), periphase (around catanaphase), paraphrase (solvent regions) and salphase (high phase contrast spots), which were similar to those reported by Jäger et al. (2004). Interestingly, the researchers, for the first time, found very poor correlation between the asphaltene content and the bee-shaped structures. Furthermore, no correlation was found between the PDM results and the SARA fractions (S-satutares, A-aromatics, R-resins, A-asphaltencs) or acid-base contents of the observed bitumen. In addition, the area ratio of the catanaphase appeared to correlate with the mass parent of Vanadium and Nickel metals for several samples. The authors also reported different microstructures of the same bitumen samples prepared by heat casting and solvent casting. Continuing their study to bitumen stiffness at low temperatures under cryogenic AFM (Masson et al., 2007), at temperatures -10°C to -30°C, they showed that not all the bitumen phases contracted equally. The reason for this different contraction could be that the catana, peri and para phases were related to the existence of domains with different glass transition temperatures. One of the key findings of this low temperature AFM study was that bitumen with multiple phases at room temperature are never entirely rigid, even when well below the glass transition temperature.

De Moraes et al. (2009), studied the high temperature behaviour of bitumen using AFM at different temperatures followed by different rest periods. The authors observed that phase contrast and topography images were highly dependent on storage time and temperature. This research group also observed the bee phase, a soft matrix phase surrounding the bee phase and a hard matrix dispersed on the soft phase, similar to findings reported by Jäger et al. (2004) and Masson et al. (2006). The authors also observed that the bee-shaped structure completely disappeared for samples at temperatures higher than 70°C and upon cooling to 66°C they began nucleating.

Concurrently, the changing microstructure after aging of bitumen was studied using AFM by Wu et al. (2009). In these studies, both neat and SBS polymer modified bitumen was aged using a pressure aging vessel (PAV). Comparing the obtained images before and after aging, the authors reported that the bee-shaped structure significantly increased after aging. To support the observed phenomena, the researchers related this to the production of asphaltene micelle structures during the aging process. These findings are consistent with the study by Zhang et al. (2011).

Tarefder et al. (2010a, 2010b) studied nanoscale characterization of asphalt materials for moisture damage and the effect of polymer modification on adhesion force using AFM. They observed wet samples always showed higher adhesive force compared with dry one. The authors concluded that the adhesion behaviour of bitumen can vary with the chemistry of the tip and functional groups on it.

Recently, Pauli et al. (2009, 2011) reported that all of these interpretations, including their previous findings were at least partially wrong and came up with a new hypothesis in which they stated that the "bees are mainly wax". To prove their hypothesis, they scanned different fractions of bitumen and found bee-shaped structures even in the maltenes which contain no asphaltenes, while the de-waxed bitumen fractions did not show any microstructures. The authors concluded that the interaction between the crystallizing paraffin waxes and the remaining bitumen are responsible for much of the microstructuring, including the well-known bee-shaped structures. They also reported sample variables such as film thickness, the solvent spin cast form, and the fact that solution concentration could also strongly influence the corresponding AFM images.

Even though different research groups concluded significantly different reasons for the structures to appear, the extensive atomic force microscopy (AFM) studies showed that bitumen has the tendency to phase separate under certain kinetic conditions and is highly dependent on its temperature history. Besides focusing on the reasons behind the microstructure growth, it is also important to relate these microstructures to the performance of bitumen.

3. Utilizing the AFM images as a basis for asphalt healing model

From Atomic Force Microscopy (AFM) scans, it was found that bitumen is not a homogeneous bulk material as microstructures are observed in almost all the bitumen (Loeber et al., 1996, 1998; Pauli et al., 2001, 2009, 2011; Jäger et al., 2004; Masson et al., 2006). All the previous studies showed that bitumen appeared to have some form of 'phase separation' dependence on its temperature history, storage time and crude source. Changing the temperature of the bitumen showed a movement of the phases and sometimes resulted in an overall homogeneous material, where the clusters seem to have 'melted' back into the matrix structure. The ability of bitumen to phase separate and redistribute its phases depends on the thermal or mechanical energy input. Recently, based on this phase separation phenomena under certain kinetic conditions, Kringos et al. (2009a) developed a healing model by assuming the bitumen matrix has two types of phases (i.e., phase α and phase β), as shown in Fig. 2.

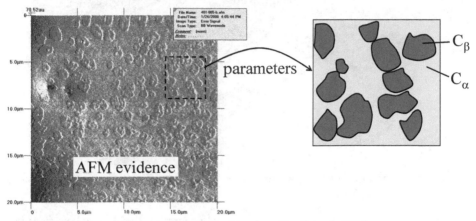

Fig. 2. AFM evidence of phase separation in bitumen (Pauli et al., 2011)

3.1 Postulated healing mechanism

From mechanical considerations, it is known that the interfaces between two materials with different stiffness properties serve as natural stress inducers. This means that when the material is exposed to mechanical and or environmental loading, these interfaces will attract high stresses and are prone to cracking. On this scale, this would result in a crazing pattern, which can be detected on a higher (macro) scale by a degradation of the mechanical properties of the material, such as the stiffness or fracture strength. If this process would continue, these micro-cracks (or crazes) would continue developing, start merging and finally form visible cracks.

A finite element simulation done by Kringos et al. (2012) demonstrated the concept of diminished response and the introduction of high stresses in an inhomogeneous material, as shown in Fig. 3, for a constant displacement imposed on a homogeneous and an inhomogeneous bitumen. The bituminous matrix is hereby simulated as a visco-elastic material and the inclusions as stiff elastic. From the deformation pattern it can be seen that the inhomogeneous material acts not only stiffer, but is no longer deforming in a smooth, uniform, manner and high stresses appear from the corners of the stiffer particles.

From the bitumen AFM scans it was shown that many bitumen sapmles, under certain circumstances, will form such inhomogeneities. If then, by changing the thermodynamic conditions of the material by inputting thermal or mechanical energy, these inclusions rearrange themselves or disappear; restoration of the mechanical properties would appear on a macro-scale. Since the phase-separation is occurring on the nano-to-micro scale, the interfaces between the clusters and the matrix could start crazing when exposed to (thermo) mechanical loading. A change of these clusters, either by rearranging themselves or by merging into the main matrix, would then lead to a memory loss of these micro-crazes and

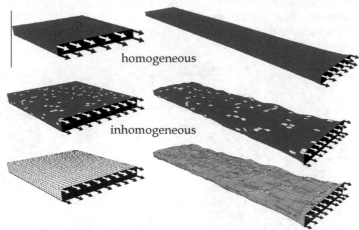

Fig. 3. Displacement and normal stress development in a homogeneous vs. inhomogeneous bitumen (Kringos et al., 2012)

the bitumen would show a restoration of its original properties, which may be referred to as physico-chemical "healing". The driving force for the rearrangement of the phases upon a changed thermodynamic state is explained in the following section.

3.2 Governing equations

To develop the governing equations for the phase-separation model, the general mass balance equation can be expressed as

$$\rho \frac{\partial_\alpha c}{\partial t} - \text{div}\left(\rho_\alpha \underset{\approx}{M} \cdot \nabla_\alpha \mu\right) = 0 \tag{1}$$

where $_\alpha\rho = \frac{_\alpha m}{V}$ is the density, $_\alpha\mu$ is the chemical potential and $_\alpha\underset{\approx}{M}$ is the diffusional mobility tensor of phase α in the material.

The chemical potential μ_α can be written as the functional derivative of the free energy Ψ :

$$_\alpha\mu = \frac{\delta\Psi}{\delta_\alpha c} \tag{2}$$

The free energy is composed of three terms

$$\Psi = \Psi_0 + \Psi_\gamma + \Psi_\varepsilon \tag{3}$$

where Ψ_0 is the configurational free energy, Ψ_γ is the Cahn Hilliard surface free energy and Ψ_ε is the strain energy.

Solving these equations, allows for the simulation of phase separation from various starting configurations. The free energy potentials and the mobility coefficient are hereby the important input parameters to the model and need to be determined for the individual bitumen samples. In Fig. 4 a comparison between AFM scans and the model is given for three different bitumen samples. More details about the analyses can be found in Kringos et.al. (2012).

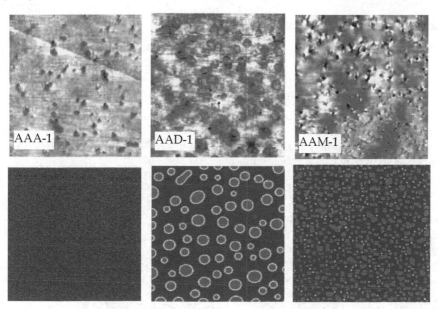

Fig. 4. Comparison between AFM scans and computed phase configuration (Kringos et al., 2012)

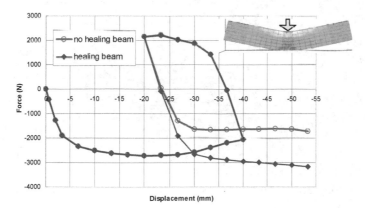

Fig. 5. Simulation of healing versus no-healing asphalt beam (Kringos et al., 2012)

The model can also be utilized in an 'upscaled' manner, in which the rearrangement of the phases, the appearance and thickness of the interfaces are linked to a healing function that becomes an intrinsic parameter that controls the evolution of dissipated energy. More details on the equations of this model and the parameters can be found from Kringos et.al.

(2009b). The force versus displacement diagram is shown for a simulation of a fatigue test using this model. It can be seen from the graphs that in the case on the 'healing beam' the material has lost its memory of the previous loading cycle, whereas in the case of the 'no-healing beam' the material is considerably weaker after the first loading cycle.

4. Healing model parameters

A parametric analysis performed by Kringos et al. (2012) investigated the effect of the various model parameters on the resulting phase configuration. In the simulations a mesh of 50 μm x 50 μm is simulated, similar to the size of the bitumen scans under the AFM. The time and space increments are normalized with respect to each other.

The initial t_0 configuration is represented by a homogenous mixture with a random perturbation, also known as a spinodal composition. Depending on the chosen parameters, different phase separation patterns will form. In the following, the influence of these parameters is shown by varying their values.

For the Cahn Hilliard parameter of $\kappa_2 = 0.0005$ and an average start concentration of 0.5 with random fluctuations of zero mean and no fluctuation greater than 0.01, results in the configurations at different time steps shown in Fig.6.

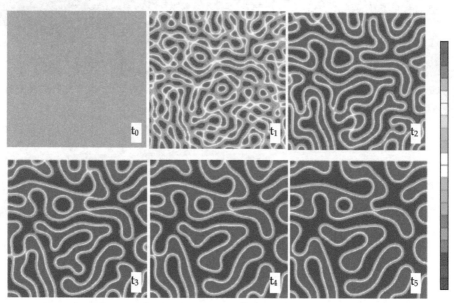

Fig. 6. Spinodal composition resulting in polymer-like phase separation (Kringos et al., 2012)

Keeping the parameters constant, but now changing the initial configuration to a maximum fluctuation of 0.05 results in a different end configuration as is shown in Fig. 7, which resembles better the phase configuration of bitumen as is seen under the AFM scans.

The Cahn Hilliard parameters are often expressed as ε which is related to the κ_2 via $\varepsilon^2 = \kappa_2$. The effect of this parameter is shown in Fig. 8.

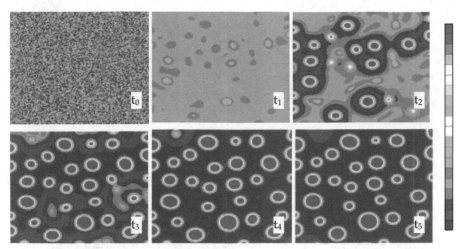

Fig. 7. Spinodal composition resulting in bee structure formation (Kringos et al., 2012)

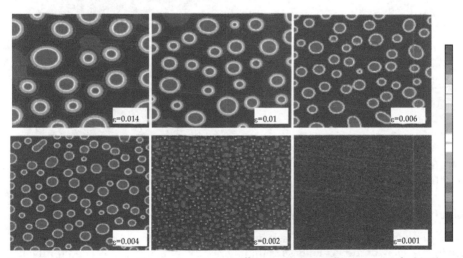

Fig. 8. End configurations with varying Cahn Hilliard parameter (Kringos et al., 2012)

Defining the matrix, the distinct phases (the bees) and the interfaces (IF) as the three fractions in the end configuration, the normalized fractions are plotted as a function of the varying Cahn Hilliard parameter, Fig. 9. As can be seen from the graph, an increased Cahn Hilliard parameter causes an increased interface (IF) fraction. Since this parameter controls the gradient energy distribution this result seems certainly logical. Interestingly thought, it can also be seen that both the matrix and the bees seem reduced. Which means that with an increased gradient distribution coefficient fewer bees are formed with thicker interfaces.

Changing the configurational free energy function whereby normalizing the energy barrier as shown in Fig. 10 (a) has the effect shown in Fig. 10 (b).

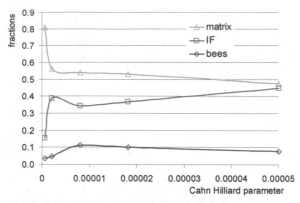

Fig. 9. Relationship between fractions and Cahn Hilliard parameter (Kringos et al., 2012)

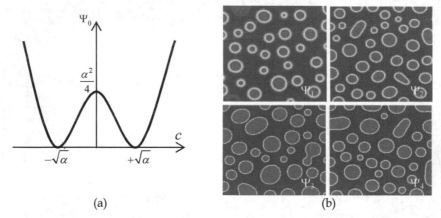

Fig. 10. (a) Normalized .. potential (b) Changing Ψ_0; α =1, 2, 4 and 6 for $\Psi_1 - \Psi_4$ (Kringos et al., 2012)

5. Mechanical properties of the phases

As was discussed in the previous sections, it is important to know the properties of the bitumen phases with respect to its surrounding matrix, which is also needed for the healing simulation in the next section, Fig. 11. The rheology at various temperatures of the individual phases turned out to be more challenging than anticipated, so to get the parameters needed for the individual phases for the healing simulation, in this section a finite element analyses is performed. In this analysis the overall bitumen properties as determined in the previous section is used.

For each case the equivalent stiffness ratio is calculated from E_{case}/E_{caseA}

where case A represents an homogeneous mesh without any phase separation. The subscript m refers to the matrix properties and the subscript b refers to the bee structures. Using the phase separated configuration as shown in Fig. 12, the properties of the matrix were varied in the analyses while assuming the bee properties remained constant at $E_b = 2E_0$ and $\eta_b = 10\eta_0$.

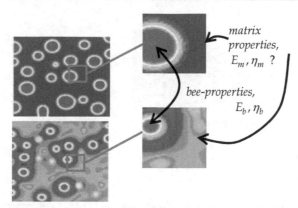

Fig. 11. Properties of bees and surrounding matrix (Kringos et al., 2012)

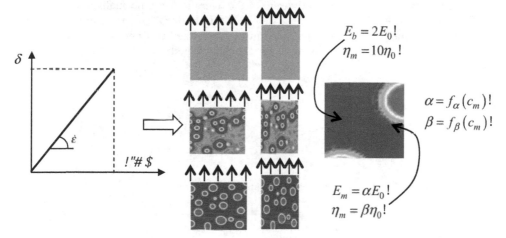

Fig. 12. Simulation of constant displacement test (Kringos et al., 2012)

From Table 1 the chosen matrix properties and the calculated equivalent stiffness at time 0.08 s is shown. In Fig. 13 the equivalent stiffness over the entire direct tension test is plotted.

Case	Configuration	E_m	η_m	Equi.stiffness ratio @ t=0.08s
A	Homogeneous	E_0	η_0	1.00
B	Phase separated	E_0	η_0	1.06
C	Phase separated	$0.5E_0$	$0.1\eta_0$	0.56
D	Phase separated	$0.5E_0$	η_0	0.82
E	Phase separated	$0.9E_0$	η_0	1.02
F	Phase separated	$0.85E_0$	η_0	1.00

Table 1. Determined parameters

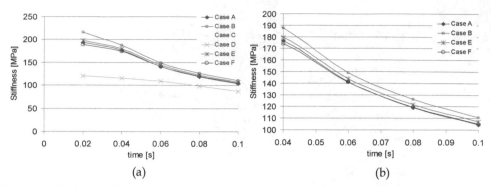

Fig. 13. Calculation of equivalent stiffness for direct tension cases (Kringos et al., 2012)

In the case of the homogeneous configuration, the overall matrix normalized concentration is 0.5 , in the case of the phase separated configuration, the minimum normalized concentration of the bees is -1 and the maximum normalized concentration of the matrix is +1, Fig. 14. Based on this, the following relationship is formed:

$$E_m = \alpha E_0 \quad ; \quad \alpha = (0.84+0.34C) -1$$

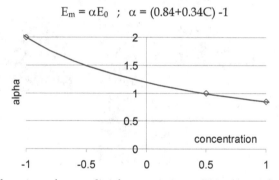

Fig. 14. Alpha as a function of normalized concentration (Kringos et al., 2012)

6. Simulation of healing in bitumen

Using the parameters as described in the previous sections, in this section the developed healing model will be demonstrated. For this, three different cases have been selected, Fig. 15.

In all cases a mesh of 50 μm x 50 μm was exposed to six displacement controlled tension and compression loading cycles. In case A, this was done continuously without any rest periods between the cycles. In case B and C, a rest period was applied after the first three loading cycles. For case B, during this rest period the material was allowed to visco-elastically unload. In case C, in addition to the visco-elastic unloading, the material was allowed to rearrange its phases, based on the PS model as was shown in the earlier sections. The initial material configuration in all simulation was the phase-separated configuration as was computed in Fig. 7. In the simulation, the mechanical properties for the matrix and the bees as derived earlier in this chapter and the energy based elasto-visco-plastic constitutive model briefly described earlier were utilized.

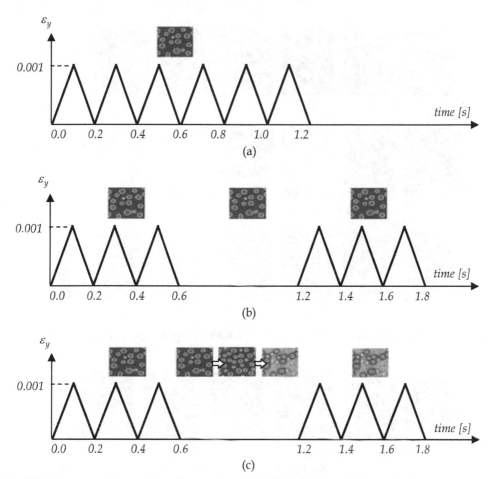

Fig. 15. Fatigue simulation (a) with no rest period (b) with VE unloading during rest period (c) with phase rearrangement during rest period (Kringos et al., 2012)

In Fig. 16 and Fig. 17 the evolution of the normal stress and damage development are shown for the simulated time steps, respectively. Damage is hereby defined as equivalent plastic strain, or permanent deformation.

What is very noticeable from the stress and damage development plots is the inhomogeneous development of the stresses and damage throughout the material. Additional it can be noticed from the damage graphs that the damage originates from the interface areas between the bee structures and the matrix.

In Fig. 17 the damage development at a chosen location in the simulated mesh is plotted as a function of the loading cycles and loading time. From these graphs it can be seen that case B has generated less damage than case A after the six loading cycles. By comparing with case C, that also included the phase rearrangement based on the kinetics of the material, it can be seen that even less damage was generated in comparison with case A and B.

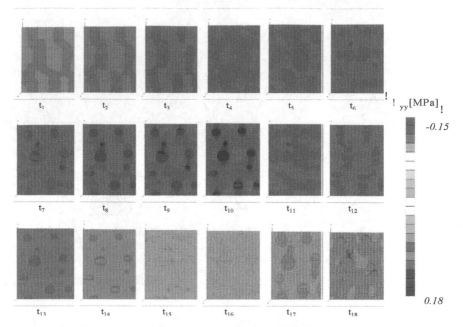

Fig. 16. Stress development during fatigue simulation (Kringos et al., 2012)

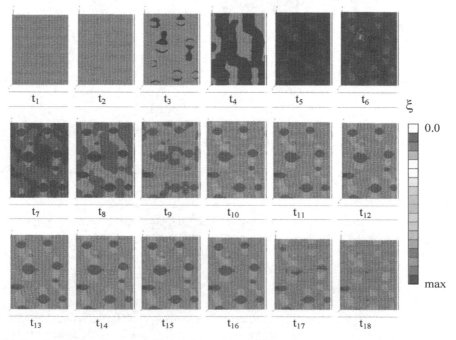

Fig. 17. Damage ξ development during fatigue simulation (Kringos et al., 2012)

Keeping in mind that these plots are made for a given location in the material, the equivalent stiffness of the overall mesh was also determined, Fig. 18. From these graphs it can be seen that case B again showed a lesser reduction of the intial stiffness after the six loading cycles as compared with case A. Case C showed even more restoration of the equivalent stiffness in comparison with case A and B.

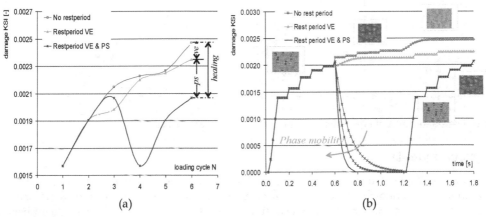

(a)   (b)

Fig. 18. Simulation of damage and healing development (Kringos et al., 2012)

From these analyses is can be concluded that the 'healing' propensity of bitumen which is generally quantified by a fatigue test, could be partly due to visco-elastic loading and partly due to the healing mechanism that was presented in this chapter. If this is indeed the case, this would have important implications for the in time development of the healing propensity of asphalt pavements in the field.

Moreover, the two contributions (visco-elasticity and phase rearrangements) are controlled by completely different parameters. Which is important to understand when one wants to optimize the healing potential of an asphaltic mixture.

For instance it can be seen from Fig. 18 (b) that by varying the phase mobility, the rate of healing can be changed. This would mean that the applied rest periods in laboratory healing tests or the breaks between traffic loading on asphalt pavements would have important effects on the healing of generated damage. It is very well possible that some bitumen have intrinsic healing capacity, but due to a relatively low mobility of the phases, this is never maximized in practice. This would mean that if bitumen producers could change something in the bitumen blend to increase the mobility of certain phases, the healing propensity could be optimized.

The material properties of the individual phases are of paramount importance for the mechanical properties and the healing capacity of the entire bitumen. In the previous section the phase scans of bitumen under the AFM were discussed. The AFM can also be utilized to determine the material properties as well as the scanned phases simultaneously. The difficulty with observing bitumen using AFM is its viscoelastic nature and its highly temperature dependent flow behavior. In the following section, the utilization of AFM to determine material properties of bituminous phases is discussed in more details.

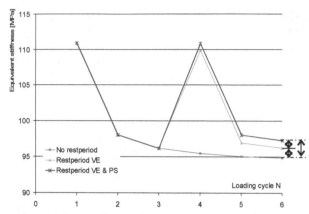

Fig. 19. Calculation of equivalent stiffness (Kringos et al., 2012)

7. Nano-indentation

Instrumented indentation testing, a test method where load and depth of indentation are continuously monitored, have become a common technique for characterizing mechanical behavior of a large variety of materials. Indentation testing is one of the few experimental techniques that can be performed at both large and small scales, thus allowing for the investigation of materials behavior across length scales from milli- to nanometers. The stress field induced by indenter in the specimen is highly localized, and the scale is proportional to the contact area. Thus by varying the indenter geometry, and size, as well as the applied load, one may obtain results representative for the material behavior on different scales. By using high-resolution testing equipment, the material properties can be measured at the micro- and nanometer level, thus allowing for a study of the variability in material properties across a specimen (Oliver & Pharr, 1992, 2003). Over the past 20 years, this technique has been used to investigate the linear elastic, viscoelastic and plastic properties of thin films, modified surfaces, individual phases in alloys and composites and other microscopic features. A detailed account of work done in this field has recently been given by (Gouldstone et al., 2007).

Recently number of attempts have been reported in the literature to use AFM in force mode to obtain elastic and viscoelastic properties of different materials, cf. e.g. (Tripathy & Berger, 2009; Gunter, 2009). AFM allows measurement of viscoelastic material properties with spatial resolution on the nanoscale; furthermore by performing the repetitive measurements at different temperatures or before and after oxidation it allows for the observance of different phases of the material. These data can be a key to understanding the governing mechanisms behind healing and ageing phenomena in bituminous binders.

The heart of any indentation testing is a method to extract the material parameters from the experimentally observed relation between indentation force F , and indentation depth, h . Modelling the indentation of viscoelastic solids thus forms the basis for analyzing the indentation experiments. Lee & Radok made substantial progress, by solving the problem of sphere indenting linear viscoelastic halfspace (Lee & Radok, 1960). Analogous results for the case of conical indenters were presented by (Graham, 1965), and for a flat punch by (Larsson & Carlsson, 1998). Huang & Lu (2007) developed a semi-empirical solution for viscoelastic indentation of the Berkovich indenter. In AFM, a probe consisting of a cone with a nominal tip radius on the order of 10 nm or higher is typically used. The accurate determination of the AFM tip shape is in fact one of the major sources of uncertainty, when performing nanoindentation testing with AFM, particularly at lower load levels. As it has been argued by many investigators, in indentation testing at micro- and nanoscales it is very difficult to make a valid assumption concerning the indenter geometry, cf. e.g. Korsunsky (2001) and Giannakopoulos (2006). Even the best attempts at preparing a perfectly round spherical or perfectly sharp conical shapes inevitably produce flattened imperfect shapes. These deviations of the indenter shapes from the assumed ideal shape will affect measurements performed at small scales. The AFM tip is normally considered to be of conical shape with round spherical tip. The contact geometry is thus dominantly controlled by the spherical indenter tip at low load levels and is switching to conical geometry at higher loads. The procedure to extract viscoelastic properties with spherical indentation is illustrated below (Larsson & Carlsson, 1998)

Using ordinary notation the stress-strain relations for linear viscoelastic solids can be formulated in relaxation form as:

$$s_{ij}(t) = \int_0^t G_1(t-\tau)\frac{d}{d\tau}e_{ij}(\tau)d\tau \tag{4a}$$

$$\sigma_{ii}(t) = 3\int_0^t G_2(t-\tau)\frac{d}{d\tau}\varepsilon_{ii}(\tau)d\tau \tag{4b}$$

where

$$s_{ij} = \sigma_{ij} - \frac{1}{3}\delta_{ij}\,\sigma_{kk}\,, \qquad e_{ij} = \varepsilon_{ij} - \frac{1}{3}\delta_{ij}\,\varepsilon_{kk} \tag{5}$$

are the deviatoric components of stress and strain. In equations (4a) and (4b), G_1 and G_2 are the so-called relaxation functions in shear and dilation, respectively. The viscoelastic Poisson's ratio, $v(t)$, is related to the relaxation functions in equations (4a) and (4b) as:

$$\overline{v(s)} = \frac{1}{s}\frac{\left(\overline{G_2(s)} - \overline{G_1(s)}\right)}{\left(2\overline{G_2(s)} - \overline{G_1(s)}\right)} \tag{6}$$

using Laplace transformed quantities according to

$$\overline{f}(s) = \int_0^\infty f(t)e^{-st}\,dt \tag{7}$$

In principle, in order to completely characterize the viscoelastic material one needs to determine two independent viscoelastic functions $G_1(t)$ and $G_2(t)$. However, these two functions cannot be determined uniquely from experimental force-displacement data. Thus in most conventional testing techniques a constant viscoelastic Poisson's ratio is assumed, and nanoindentation measures only the relaxation compliance in shear, cf. e.g. Giannakopoulos (2006) and Jäger et al. (2007). The shear relaxation function G_1 is expressed then as Prony series and the introduced constants are then determined in order to best fit the experimental results.

The relaxation test with AFM is performed by programming the tip to indent the specimen to a specified depth, $h(t)$, and then to hold the penetration depth constant for a specified time, i.e. $h(t) = h_0 H(t)$, $H(t)$ being the Heaviside function. Provided that the indenter is stiff as compared to the sample, the relation between G_1 and the indenter load measured as a function of time, $P(t)$, is given than as:

$$G_1(t) = \frac{6R(1-v)P(t)}{8a_0^3} \tag{8}$$

where R is the radius of curvature of the spherical indenter tip and a_0 is the maximum radius of the contact area defined as:

$$a_0^2 = (Dh_0)/2 \tag{9}$$

Equation (8) may be used to obtain shear relaxation modulus, provided that constant Poisson's ration is known, the material response is linear and the contact geometry is spherical. Equation (9) provides a simple way to check the linearity of the material response. As it has been show by Larsson & Carlsson (1998), the size of the residual impression in the specimens surface is very close to the maximum size of the contact area. One may thus follow AFM indentation testing with AFM scanning in tapping mode to measure size of the residual impression and compare it with the maximum contact radius predicted by equation (9). The discrepancy between the measured and predicted contact radii would indicate the presence of non-linear material effects. The assumption of the constant Poisson's ratio may not be satisfactory for characterization of bituminous binders, as volumetric material response may also exhibit spatial variation. Researchers have been attempting to address this issue and distinguish dilation and shear compliances by using secondary sensors to measure circumferential strain at a small distance from the contact area, e.g. Larsson & Carlsson (1998) or by comparing force-displacement relations for indenters with different shapes, e.g. Huang & Lu (2007). In particular, Larsson & Carlsson (1998) derived the following relations between $G_1(t)$, $G_2(t)$ and $P(t), \varepsilon_\theta$:

$$\int_0^t G_1(\tau)d\tau = \frac{6R}{8a_0^3}\int_0^t P(t-\tau)(1-v(\tau))d\tau \tag{10}$$

$$\int_0^t \varepsilon_\theta(r,t-\tau)(1-v(\tau))d\tau = -\frac{4a_0^3}{6\pi Rr^2}\int_0^t (1-2v(\tau))d\tau \tag{11}$$

Equations (10) and (11) are uncoupled Fredholm integral equations of first and second kind respectively. The technique required to solve such a system of equations numerically is well documented, cf. e.g. Andersson & Nilsson (1995).

Huang & Lu (2007) suggested to use the semi-empirical force-displacement relations for the Berkovich indenter along with the analytical solution for the spherical indenters as two independent equations for establishing $G_1(t)$ and $G_2(t)$. The relaxation functions may be then expressed as Prony series and their coefficients may be found as the best fit of the experimental data.

It has to be pointed out however that both of these approaches appear to be questionable when applied to AFM indentation testing: instrumenting the specimens surface with additional sensors appears not to be practically feasible at the length scale in question; at the same time the approach presented in Huang & Lu (2007) would require AFM tips with two distinct well defined shapes. Furthermore, the possibility to accurately measure linear viscoelastic material behavior with sharp indenters, such as Berkovich or conical, has been questioned by several researches, cf. e.g. Vanlandingham et al. (2005). The reason is that sharp indenters inevitably introduce intense strains local to the indenter tip, thus making the assumption of linearity invalid.

It may be concluded from the above that using AFM in force mode provides a useful tool to obtain at least qualitative information regarding the bitumen properties at nano-scale and their evolution with temperature and oxidation. The use of AFM to obtain absolute quantitative estimates of viscoelastic properties of the bituminous binders appears however questionable at the present moment. One way to proceed is to complement AFM measurements by instrumented indentation measurements at microscale and macroscale, in order to obtain better initial quantitative estimates of the volumetric and shear relaxation functions.

8. Discussion and recommendations

From simulations with the developed healing model, briefly described in the previous sections, it is evident that many different bitumen phase configurations can be formed from an initially homogeneous configuration. The initial configuration and the smallest local dispersions of the material can thereby have an important effect, as does the configurational and surface free energy potentials. The rate at which these changes occur is of great importance for the practical implications of this phenomenon relates to rest periods in the laboratory to assess the healing rates and the actual healing ability of the asphalt pavement. Mobility of the bitumen is thereby an important characteristic of the bitumen that affects the healing potential and should be determined. It can therefore be expected that enhancing the mobility of the bituminous phases will increase the healing rates and thus the pavement life time.

The configuration free energy potential of bitumen can be determined from the configuration free energy curves of the individual phases by assuming that in a certain composition only the phase with the lowest energy will exist. Equilibrium is then found when the chemical potential is homogeneous throughout the material. This does not, however, mean that the mass fraction has to be homogeneous. Inhomogeneous distribution of phase affects the ability for bitumen to transmit stresses through its matrix. Having knowledge about the bitumen configurational free energy potentials can therefore contribute to better predictions of the in-time evolution of its ability to transmit stresses and reduce damage propagation. Optimizing the bitumen phase diagrams to promote the resilience for stress-endurance will therefore enhance the lifetime of the overall pavement. This means that molecular volumes, densities and solubility of the bitumen phases should be determined.

The surface free energy potential accounts for energy due to interfaces between the distinct phases. For diffuse interfaces this means that the free energy is dependent on both the local composition as well as its surroundings. In the developed healing model, the Cahn Hilliard approach is taken in which the generation and thicknesses of the interfaces depends on a gradient energy coefficient. Insight into the gradient energy coefficients of various bitumen would enable more accurate predictions into the evolution of damage and healing potential and would allow for an optimization of the long-term behaviour.

With the advent of more powerful experimental tools with better controlled environmental conditions, it is foreseen that future research will be able to accurately and more easily determine these parameter for bitumen. Research will continue to develop the predictive models and more insight into the fundamental parameters that influence the healing and damage resistance of bitumen will be generated. It can be envisioned that from this research, detailed databases with the most commonly used bitumen will be developed that can be included in future asphalt mixture and pavement design. Also, more detailed test procedures that would enable researchers and pavement engineers to measure these parameters for their own material will become available in the coming years.

9. References

Andersson, M. & Nilsson, F. (1995). A perturbation method used for static contact and low velocity impact. *International Journal of Impact Engineering*, Vol. 16, No. 5, pp.759-775, ISSN 0734-743X

Baker, H. (1742). *The Microscope Made Easy*, Science Heritage Ltd, ISBN 0940095033, Lincolnwood,IL. 1987 reprint

de Moraes, M.B., Pereira, R.B., Simao, R.A., Leite, L.F.M. (2010). High temperature AFM study of CAP 30/45 pen grade bitumen. *Journal of Microscopy*, Vol. 239, No. 1, pp. 46–53, ISSN 1365-2818

Giannnakopoulos, A. E. (2006). Elastic and Viscoelastic Indentation of Flat Surfaces by Pyramid Indentors. *Journal of Mechanics and Physics of Solids*, Vol. 54, No. 7, pp. 1305-1332, ISSN 0022-5096

Gouldstone, A., Chollacoop, N., Dao, M., Li, J., Minor, A. & Shen, Y.-L. (2007). Indentation Across Size Scales and Disciplines: Recent developments in experimentation and modeling. *Acta Materialia*, Vol. 55, No. 142, pp. 4015- 4039, ISSN 1359-6454

Graham, G. A. C. (1965). The contact problem in the linear theory of viscoelasticity. *International Journal of Engineering Sciences*, Vol. 3, No. 1, pp. 27-46, ISSN 0020-7225

Gunter, M. (2009). AFM Nanoindentation of Viscoelastic Materials with Large End-Radius Probes. *Journal of Polymer Science: Part B: Polymer Physics*, Vol. 47, No. 16, pp. 1573-1587, ISSN 0887-6266

Hesp, S.A.M., Iliuta, S. & Shirokoff, J.W. (2007). Reversible aging in asphalt binders. *Energy & fuels*, Vol. 21, No. 2, pp. 1112-1121, ISSN 0887-0624

Huang, G. & Lu, H. (2007). Measurements of Two independent Viscoelastic Functions by Nanoindentation. *Experimental Mechanics*, Vol. 47, No. 1, pp. 87-98, ISSN 0014-4851

Jäger, A., Lackner, R., Eisenmenger-Sittner, Ch. & Blab, R. (2004). Identification of four material phases in bitumen by atomic force microscopy. *Road Materials and Pavement Design*. Vol. 5, pp. 9–24, ISSN 1468-0629

Jäger A., Lackner, R. & Stangl, K. (2007). Microscale Characterization of Bitumen–Back-Analysis of Viscoelastic Properties by Means of Nanoindentation. *International Journal of Materials Research*, Vol. 98, No. 5, pp. 404-413, ISSN 1862-5282

Korsunsky, A.M. (2001). The Influence of Punch Blunting on the Elastic Indentation Response. Journal of Strain Analysis, Vol. 36, No. 4, pp. 391-400, ISSN 0309-3247

Kringos, N., Schmets, A., Pauli, T. & Scarpas, T. (2009a). A Finite Element Based Chemo-Mechanics Model to Simulate Healing of Bituminous Materials, *Proceedings of the international workshop on chemo-mechanics of bituminous materials*, pp. 69-75, ISBN 978-94-90284-04-6, Delf, The Netherlands, June 9-11, 2009

Kringos, N., Scarpas, A., Pauli, T. & Robertson, R. (2009b). A thermodynamic approach to healing in bitumen. In: *Advanced testing and characterization of bituminous materials*, Loizos, A., Partl, M.N. & Scarpas, T., pp. 123- 128, Taylor & Francis, ISBN 978-0-415-55854-9, London

Kringos, N., Pauli, T., Schmets, A. & Scarpas, T. (2012). Demonstration of a New Computational Model to Simulate Healing in Bitumen. *Journal of the Association of Asphult Paving Technologists*, Under review

Larsson, P.-L. & Carlsson, S. (1998). On Microindentation of Viscoelastic Polymers. *Polymer Testing*, Vol. 17, No. 1, pp. 49-75, ISSN 0142-9418

Lee, E. H. & Radok, J. R. M. (1960). The contact problems for viscoelastic bodies. Journal of Applied Mechanics, Vol. 27, No. 3, pp. 438-444, ISSN 0021-8944

Lesueur, D., Gerard, J.-F., Claudy, P., Létoffé, J.-M., Planche, J.-P. & Martin, D. (1996). A structure-related model to describe asphalt linear viscoelasticity. *Journal of Rheology*, Vol. 40, No. 5, pp. 813–836, ISSN 0148-6055

Loeber, L., Sutton, O., Morel, J., Valleton, J.-M. & Muller, G. (1996). New direct observations of asphalts and asphalt binders by scanning electron microscopy and atomic force microscopy. *Journal of Microscopy*, Vol. 182, No. 1, pp. 32–39, ISSN 1365-2818

Loeber, L., Muller, G., Morel, J. & Sutton, O. (1998). Bitumen in colloidal science: a chemical, structural and rheological approach. *Fuel*, Vol. 77, No. 13, pp. 1443–1450, ISSN 0016-2361

Lu, X. & Redelius, P. (2006). Compositional and structural characterization of waxes isolated from bitumens. *Energy & fuels*, Vol. 20, No. 2, pp. 653-660, ISSN 0887-0624

Oliver, W. C. & Pharr, G. M. (1992). An Improved Technique For Determining Hardness and Elastic-Modulus Using Load and Displacement Sensing Indentation Experiments. *Journal of Materials Research*, Vol. 7, No. 6, pp. 1564-1583, ISSN 0884-2914

Oliver, W. C. & Pharr, G. M. (2003). Measurement of Hardness and Elastic Modulus by Instrumented Indentation: Advances in Understanding and Refinements to Methodology. *Journal of Materials Research*, Vol. 19, No. 1, pp. 3-20, ISSN 0884-2914

Pauli, A.T., Branthaver, J.F., Robertson, R.E., Grimes, W. and Eggleston, C.M. (2001). Atomic force microscopy investigation of SHRP asphalts. *ACS division of fuel chemistry preprints*, Vol. 46, No. 2, pp. 104-110

Pauli, A.T. & Grimes, W. (2003). Surface morphological stability modeling of SHRP asphalts. *ACS division of fuel chemistry preprints*, Vol. 48, No. 1, pp. 19-23

Pauli, A.T., Grimes, R.W., Beemer, A.G., Miller, J.J., Beiswenger, J.D. & Branthaver, J.F. (2009). Studies if the Physico-Chemical Nature of the SHRP Asphalts: PART-I & II, *Proceedings of the international workshop on chemo-mechanics of bituminous materials*, pp. 25-30 & 49-53, ISBN 978-94-90284-04-6, Delf, The Netherlands, June 9-11, 2009

Pauli, A.T., Grimes, R.W., Beemer, A.G., Turner, T.F. & Branthaver, J.F. (2011). Morphology of asphalts, asphalt fractions and model wax-doped asphalts studied by atomic force microscopy. *International Journal of Pavement Engineering*, Vol. 12, No. 4, pp. 291-309, ISSN 1029-8436

Redelius, P. (2009). Chemistry of Bitumen or Asphaltenes Where are you ? *Proceedings of the international workshop on chemo-mechanics of bituminous materials*, pp. 31-35, ISBN 978-94-90284-04-6, Delf, The Netherlands, June 9-11, 2009

Tarefder, R.A. & Arifuzzaman, M. (2010a). Nanoscale Evaluation of Moisture Damage in Polymer Modified Asphalts. *ASCE Journal of Materials in Civil Engineering*, Vol. 22, No. 7, pp. 714–725, ISSN 0899-1561

Tarefder, R.A., Zaman, A. & Uddin, W. (2010b). Determining Hardness and Elastic Modulus of Asphalt by Nanoindentation. *ASCE International Journal of Geomechanics*, Vol. 10, No. 3, pp. 106–116, ISSN 1532-3641

Tripathy, S. & Berger, E.J. (2009). Measuring Viscoelasticity of Soft Samples Using Atomic Force Microscopy. *Journal of Biomechanical Engineering-Transactions of the ASME*, Vol. 131, No. 9, pp. 094507, ISSN 0148-0731

Vanlandingham, M. R., Chang, N.-K., Drzal, P.L., White, C. C. & Chang, S.-H. (2005). Viscoelastic Characterization of Polymers Using Instrumented Indentation. I. Quasi-Static Testing. *Journal of Polymer Science: Part B: Polymer Physics*, Vol. 43, No. 14, pp.1794-1811, ISSN 0887-6266

Wu, S.P., Pang, L., Mo, L.-T., Chen, Y.-C. & Zhu, G.-J. (2009). Influence of aging on the evolution of structure, morphology and rheology of base and SBS modified bitumen. *Construction and Building Materials*, Vol. 23, No. 2, pp. 1005–1010, ISSN 0950-0618

Zhang, H., Wang, H. & Yu, J. (2011). Effect of aging on morphology of organo-montmorillonite modified bitumen by atomic force microscopy. *Journal of Microscopy*, Vol. 242, No. 1, pp. 37–45, ISSN 1365-2818

11

Atomic Force Microscopy – For Investigating Surface Treatment of Textile Fibers

Nemeshwaree Behary and Anne Perwuelz
ENSAIT-GEMTEX : ENSAIT, GEMTEX, Roubaix
Univ Lille Nord de France, USTL, F-59655, Villeneuve d'Ascq Cedex
France

1. Introduction

Textile fibers either natural or man-made (biodegradable and/or non biodegradable) are being increasingly used in non-traditional sectors such as technical textiles (automotive applications), medical textiles (e.g., implants, hygiene materials), geotextiles (reinforcement of embankments), agrotextiles (textiles for crop protection), and protective clothing (e.g., heat and radiation protection for fire fighter, bulletproof vests, and spacesuits).Textile structures (roving, knitted, woven or non woven) are also being increasingly used in textile reinforced composites.

Surface treatments of textile fibers, yarns or fabrics play an important role in their processing and end-use. The AFM-Atomic Force Microscopy seems a very valuable tool for investigating the effect of different fiber surface treatments and their impact on the final textile material properties. The AFM probe has been used to understand the frictional behaviour of sized glass fibers, and to study the impact of an air-atmospheric plasma treatment on polyethylene terephthalate fabrics

PART I: Use of AFM/LFM tool for friction analysis of sized glass fibers

1. Introduction

Glass fibers, generally used to reinforce composite materials, readily suffer abrasion damage due to friction when glass filaments slide against each other. In manufacturing, glass fibers are coated with a size consisting of a coupling agent, a lubricant, a film former and other additives. While the coupling agent is used to increase adhesion between the fibers and the matrix, in glass fiber reinforced composite materials (P.Plueddeman, 1982), the complete size should improve the frictional performance of contacting fibers surfaces during their processing and their uses (e.g.: spinning, weaving...). With the increasing demand for good sizing agents which have low friction values, it is important to study the frictional behavior of sized fibers.

1.1 Theoretical background of frictional properties of fibers

The frictional properties of polymer fibers deviate from the classical Amonton law. For polymers and fibers which are viscoelastic materials, the friction coefficient 'μ' depends on the temperature, speed, and more specifically on the applied load. The most striking characteristic is that the coefficient of friction 'μ' increases as the load 'N' is diminished, and at light loads, it can be very high. Indeed, other than deformation at asperities observed in the fiber, the radius of curvature of the fiber surface is an important parameter in determining friction. Gupta described the friction of fibers in empirical terms (Gupta, 1992):

$$F_f = aN^n \tag{1}$$

with

$$\mu = aN^{n-1} \tag{2}$$

where 'a' and 'n' are constants, and $0<n<1$.

The friction coefficient 'μ' reaches its maximum value at zero load where the friction forces due to adhesive forces are present. In order to initiate motion during sliding, a certain force called static friction force 'F_s' must be applied, which is greater than the kinetic friction force 'F_k' needed to maintain the sliding. This may result in stick-slip motion when one fiber slides on another one, and this intermittent motion can be generated even in the presence of the size (Gupta, 1985, 1992). Friction force is reported to be at its highest at the stick phase (Israelachvili, 1993).

1.2 Static friction force and its relation to surface contact area and surface forces

It is generally accepted that static friction depends on adhesion forces, and recently Israelachvili (Israelachvili, 1993) has demonstrated that the frictional energy required during the static phase is used partly to overcome adhesion and partly to overcome the load. At very small loads, as in our case (nanonewtons for the LFM and micronewtons for the microbalance), the influence of load is small but adhesion forces (F_{ad}) are present. These forces depend on surface energy and on the apparent contact area. Surface energies are defined constants of materials, while the theoretical contact areas can be determined by contact mechanics theories. The first contact theory elaborated by Hertz concerns elastic solids for which the contact area for two contacting spheres, is a circle of radius 'a'. However in presence of adhesive forces, the contact areas may be altered and the two contact mechanics theories JKR (Johnson-Kendall-Roberts) (Johnson, 1971) and DMT (Derjaguin-Muller-Toporov) (Derjaguin, 1975) determine the real surface of contact.

The nature of fiber friction is known to be complex, and adequate theory describing it, is not available, though models have been provided to explain frictional behavior of fibrous materials (Gupta, 1992, 1993). The aim of this work was to use lateral force microscopy to investigate friction at nanoscale level and to relate the results to those obtained at a higher scale that is, during fiber-fiber friction, using an electronic microbalance.

2. Experimental

2.1 Friction force measurements by Lateral force microscopy

The atomic force microscopy/lateral force microscopy (AFM/LFM) invented by Bining & al. in 1986 (Binning, 1986) was used to obtain topographic image in the AFM mode, and lateral forces between a $Si_3 N_4$ tip and a fiber surface in the LFM (lateral force microscopy) mode

(Meyer and Amer, 1990). Extensive reviews give detailed description of this apparatus and therefore only a brief explanation of the method will be described in this paper.

AFM imaging was achieved in air under atmospheric conditions with a commercial scanning probe «Nanoscope III, from Digital inc.», in the contact mode at a constant force. The AFM measures the vertical deflection of a cantilever to which is fixed a microtip which scans the sample surface (see Fig. 1). In the contact mode, the normal deflection (due to intermolecular forces) of the cantilever is dependent on the distance between the tip and the sample, such that $F_z = k. z$, where : F_z= deflection force of the cantilever, k= spring constant of the cantilever, and z= the cantilever's deflection.

The fiber sample is placed onto a piezodrive, while the tip is in a fixed position. The normal deflection of the cantilever is monitored through the displacement of a laser beam reflected off the cantilever onto a segmented detector, during the scanning of the sample in x and y directions (Fig. 1). Any bending of the cantilever due to bumps or grooves on the sample surface induces an intensity difference between the lower and upper part of the segmented photodetector, and this in turn provides an error signal for the feedback of the piezo so as to maintain a constant preset force on the cantilever. The displacement of the piezodrive allows the reconstitution of a topographic image of the scanned surface. The lateral force images were obtained by measuring the torque imposed on the cantilever by the tip as a result of tangential forces experienced by the tip, when the sample is moved underneath. To measure normal and lateral forces simultaneously a four quadrant photodiode is used (Fig. 1). The normal bending of the cantilever is measured by the intensity difference ($I_{1+2} - I_{3+4}$) of the upper and lower segments of the diode, while the signal difference of the left and the right segments ($I_{1+3} - I_{2+4}$) provides torsional information.

2.1.2 Topographical and LFM images

Topographical 3-dimensional as well as scope mode images were obtained in the AFM and LFM modes. Scope mode images give trace and retrace profiles of the topography and of LFM signals in real time when the sample goes forward and backward underneath the tip (see Fig. 2). Lateral force signals are influenced by the surface topography (slopes of a surface feature for example). For a surface with friction as well as with corrugations (see Fig. 2A and B), like the glass fibers used in this study, the normal reacting force (N) as well the frictional force (F_f) will have nonzero x, y, and z components, that is,

$$N = N_x + N_y + N_z$$

$$F_f = F_{fx} + F_{fy} + F_{fz}$$

In other words, the measured lateral force depends on the local slope as well and not only on frictional forces. The frictional force (F_f) is, of course, tangent to the slope (see Fig. 2B) and acts in the opposite direction to the scan direction, and, therefore, by the reversal of the scan direction, each of the components of the frictional force (F_f) changes sign, while those due to the normal force do not. The difference between the forward and the reverse scans in the LFM scope mode gives twice the average friction force (Overney and Meyer, 1993) Baselt and Baldeschwielder, 1992). Full quantification of the frictional force is not yet possible, so scanning was carried out 90° (Fig. 1) to cancel the y and z components of the friction force. Friction forces in volts were converted to Newtons by determining the lateral sensibility of

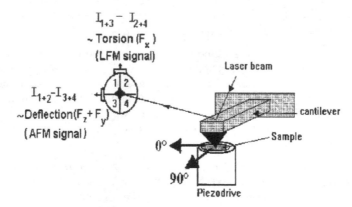

Fig. 1. Principle of simultaneous measurement of the normal and lateral forces; two scanning directions are possible (0° and 90°).

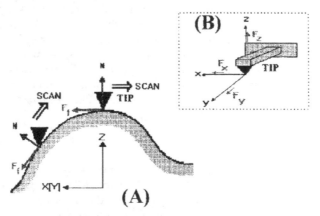

Fig. 2. (A) Normal reacting force (N) as well frictional force (F_f) acting on a surface with corrugations; (B) schematic presentation of the X, Y, and Z components of the forces acting at the top of the cantilever

the apparatus which was found to be 40 nN/V. Friction coefficients were obtained by dividing the friction force by the normal applied force.

2.1.3 Sample preparation

Sized glass filaments were provided by Owens Corning in the form of multifilament rovings. One or several of the filaments which are cylindrical in shape, were fixed onto a double-face Scotch tape, perpendicularly to the scan direction (90°), so as to measure both topographical and frictional data. This perpendicular position enables one to see in the scope mode the exact position of the fiber with respect to the tip point. As the maximum of the piezodrive in the z direction is limited to 5.9 μm, the fiber cannot be scanned wholly. So, only the most elevated part of the fiber, where the slope (or curvature) is minimum, was rastered. We shifted from a 6 x 6 to a 3 x 3μm² surface by zooming the top of the fiber, in the image mode, in real-time.

Samples were rastered at a constant force between 20 and 90 nN using a "J" head with scan area of 130 x 130 μm2 and conical-shaped ultralevers made from silicon nitride attached to the cantilever (180 μm long) with a spring constant k = 0.06 N/m. Topographic and lateral force images were obtained at a scan rate of 1.12 Hz with 512 samples per area scanned.

2.2 Fiber-Fiber friction force measurements using an electronic microbalance

An experimental device was set up using an electronic microbalance to measure fiber-fiber friction. An electronic microbalance having a sensitivity of 10^{-8} N was used since friction forces measured for two monofilaments of diameter 11 μm sliding against each other are of the order of a few micronewtons. The microbalance was equipped with an inner chamber to protect test samples from contaminants and to perform experiments in specific environmental conditions (humidity of 45%, and 25°C). The data acquisition and control station allowed force measurements at the sample weighing position, as a function of the vertical displacement of the platform.

During friction force measurements the vertical fiber was connected to the microbalance at the sample weighing position (see Fig. 3 A) and a load of 1 mg was applied at the other end of this fiber so as to maintain it in a vertical position. The horizontal fiber was fixed to a metallic holder, and brought into contact with the vertical fiber until the deflection angle of the vertical fiber was equal to 2.3°, which is also equal to the wrap angle of the horizontal fiber by the vertical one. The platform was then raised at a fixed speed of 2 μm/sec which induced sliding of one glass fiber against the other one. The force was recorded by the balance as a function of the vertical displacement of the platform. The total sliding distance along the vertical fiber was fixed to 2 mm.

2.2.1 Typical curve obtained

The experimental curve stick-slip curve obtained is quite irregular (see Fig. 3B). A detailed statistical analysis of the force curves was carried out using a computer program.

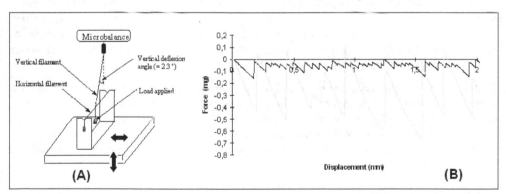

Fig. 3. (A) : Experimental set-up for friction measurement of two crossed glass fibers : the deflection angle of the vertical fiber connected to the microbalance is 2.3°, the horizontal fiber is fixed to a fiber-holder B) Stick-slip friction curve profile of fiber A (in dark line) and of fiber E (in grey line); the change in force is plotted against the vertical displacement of the horizontal fiber.

Static friction force values were obtained from the minimum values of the curve because, at the start, the microbalance platform was constrained to move upwards, and consequently static friction force values measured had negative values. So the signs of the friction force values were inverted. Moreover, before the vertical displacement of the platform, the balance was set to zero, while in practice, there was a load N at the extreme end of the vertical fiber. The real static (or dynamic) friction force measured was then (-Fs+ N).

2.2.2 Static friction coefficients

Capstan method generally applied for yarn-yarn friction measured by the F-meter was used to calculate the static friction coefficients during fiber-fiber motion (Gupta, 1993):

$$\mu = \frac{Ln\frac{T_2}{T_1}}{\alpha} \tag{3}$$

'T_1' is the initial tension, that is the load applied to the vertical fiber + the fiber weight

'T_2' is the real static force measured. (-F_s+ N). The wrap angle 'α' is equal to the deflection angle 'θ' of the vertical fiber for small values of deflection angle (see Fig. 3A). The deflection angle of the vertical fiber was calculated for each vertical displacement 'Z' of the horizontal fiber before calculating the friction coefficient

3. Results and discussions

3.1 Results of LFM analysis of fibers

Images obtained in the AFM mode (topography) and in the LFM mode are presented in figures 4, 6, 7 & 8. The difference between the LFM backward and forward signals gives twice the friction force in volts. Friction coefficients were calculated by dividing the friction force (in nN) by the normal applied force (in nN), the latter being calculated and calibrated by using the contact force curve profile (see Fig. 5A), as described earlier. For each fiber, numerous tests were performed and the friction coefficient values were found to be reproducible when tests were carried out on different regions of a filament as well as on different filaments of the same glass fiber.

The friction coefficient values were found to be relatively small and varied from 0.01 to 2. These are relevant values, especially when we compare them to those obtained by other authors (Bhushan, 1985) who worked on polymeric films and other surfaces. As compared to film-film friction coefficient values for PET for example which are around 0.66-0.77, nanoscale friction coefficients of PET films measured by AFM/LFM are relatively small and are around 0.02-0.06.

3.1.1 Desized glass fiber

Glass fiber without size was equally analyzed by AFM/LFM. A desizing procedure was established, which consisted in heating the sized fiber at 600°C for 24 h so as to completely destroy the organic size. The topographical image (Fig. 4) shows that the desized fiber surface has a friction coefficient µ= 0.04 which stays constant all throughout the fiber surface. Fig. 5C shows that there is a great attraction of the tip by the sample before the former may be in contact with the fiber. This phenomenon may be explained by a higher surface energy of the clean desized glass fiber which is going to attract water molecules of the air very rapidly. Indeed, molecules of water form a film of water at the fiber surface, and this acts as a lubricant—hence, the very weak value of the friction coefficient.

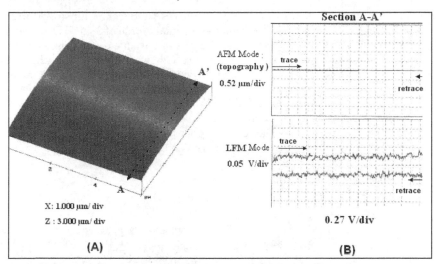

Fig. 4. Desized glass fiber: (A) topographic image; (B) scope-mode forward and backward scanned AFM and LFM signals of section A–A'.

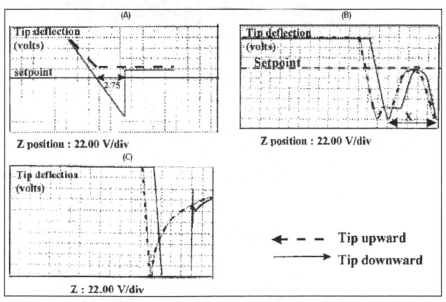

Fig. 5. (A) A normal contact force profile, (B) Contact force profile during scanning of fiber E and of (C) the desized glass fiber

3.1.2 Sized glass fibers

All sized glass fibers have very heterogeneous surfaces compared to the plain surface of the cleaned glass fiber.

3.1.2.a Fiber A

Typical images of fiber A, obtained in the AFM and LFM modes are illustrated in Fig. 6A and 6B, respectively. These have been obtained at a contact force of 38 nN. The topographic image in Fig. 6A, shows randomly distributed bumps of variable dimensions. The forward and backward scanned images in the LFM mode presented in Fig. 6 B, show that contrasts on the bumps are reversed when the scanning direction is changed. The scope mode AFM and LFM signals of the sections A-A' and B-B' corresponding to extremities of the topographic image of Fig. 6A are illustrated in Fig. 6C and 6D respectively. The scope mode LFM signals of the bumps reveal a friction force (F_f) of ~0.035 Volts (indicated by full lines), that is a friction coefficient μ ~0.04. The remaining surface (indicated by dashed lines) has a friction force two to three times greater, that is a friction coefficient μ~0.12.

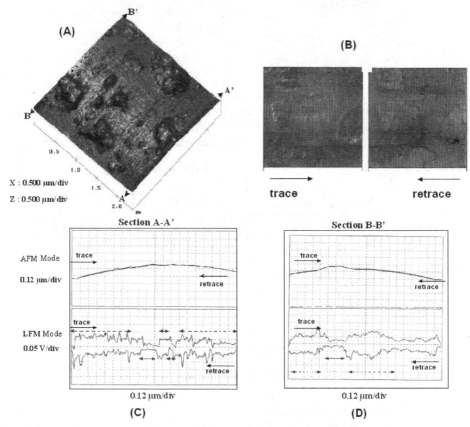

Fig. 6. Fiber A: (A) topographic image; (B) forward and backward scanned LFM images; (C) scope-mode forward and backward scanned AFM and LFM signals of section A–A' and of section B–B' (D)

Furthermore, in the LFM mode, small fluctuations of the backward and forward signals were observed. Friction forces were determined by considering the average value of signals, only.

The glass fiber with the size A presents therefore, a physically and chemically heterogeneous surface, with bumps having a friction coefficient μ~0.04 while the overall surface has a friction coefficient μ~0.12. This would mean that during coating of glass fiber A by the starch size, the different constituents of the size are not distributed in a homogeneous manner.

3.1.2.b Fiber E

When the topographic images were realized at a contact force greater than 70 nN, surface damage due to plowing of the size E by the AFM tip, was observed. A typical example of such topographic images is illustrated in Fig. 7A.

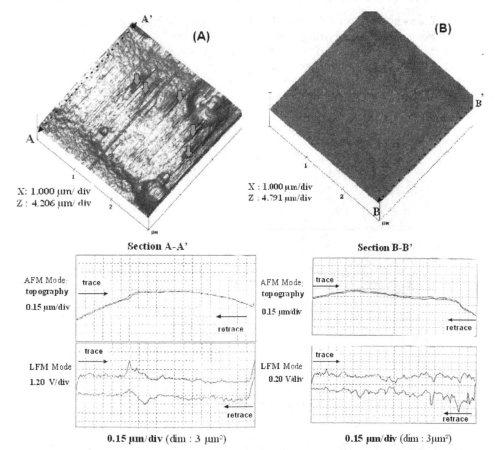

Fig. 7. Topographic image of fiber E, E4 and scope-mode forward and backward scanned AFM and LFM signals of section A–A'for fiber E, and of section B–B'for fiber E4. (plowing positions of fiber E are indicated by arrows)

Moreover, the contact force profile which gives the deflection signal as a function of the distance between the AFM tip and the fiber sample (as the piezodrive moves upwards and then downwards), shows that the AFM tip remained stuck to the size, and it was difficult to pull-off the tip from the fiber surface (Fig. 5B).

An example of topographic image of another fiber taken from the same roving of fiber E4, where plowing did not occur, is shown in Fig. 7B. The latter was made at a contact force of 62 nN and the contact force curve profile was normal. The difference between the forward and backward LFM signals revealed a constant friction coefficient of 0.2 throughout the fiber surface. This value is greater than that of the starch size on fiber A, for which the maximum friction coefficient is around 0.12.

The fact that some fibers are readily plowed by the AFM tip while others are not, may indicate a difference in the degree of cross-linking of the epoxy film-former of the size on fiber E, among fibers of a same roving.

When fiber E was annealed at 100°C for 60 h, the topographic image realized at 46 nN (Fig. 8A) shows no presence of plowing, and the contact force calibration curve profile was normal (see Fig. 5A). Nevertheless, it reveals an aggregation of matter with the formation of small blisters (indicated by arrows) of diameters of approximately 0.15 μm. LFM signals of a 1-μm² scanned area (Fig.8B) of a blister (full line) reveal a friction coefficient of μ= 0.03 which is smaller than that of the general surface of μ= 0.09.

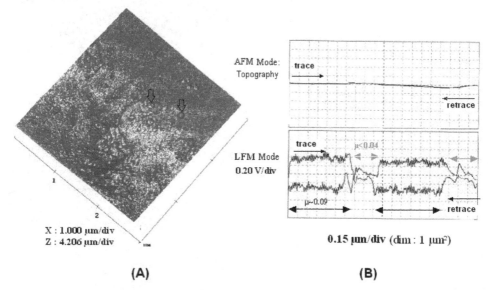

X : 1.000 μm/div
Z : 4.206 μm/div

(A)

AFM Mode: Topography

LFM Mode 0.20 V/div

0.15 μm/div (dim : 1 μm²)

(B)

Fig. 8. Annealed fiber E41: (A) topographic image; (B) scope-mode forward and backward scanned AFM and LFM signals.

According to the observations made on the annealed glass fiber E4, in both the AFM and LFM modes, it can be said that the increased crosslinking of the epoxy resin by annealing leads to a sized surface having a higher surface Young's modulus. The size is therefore less susceptible to be plowed by the AFM tip. One can also emit the hypothesis that during an increased crosslinking of the epoxy film former the lubricant is expurgated off onto the external surface in the form of blisters. This would have, consequently, the effect of decreasing the coefficient of friction ($\mu = 0.03$).

3.2 Fiber-fiber friction force measurements by the electronic microbalance

Results for measurements carried out under the following fixed conditions of constant speed (2 µm/sec), load (1 mg), relative humidity (45%), and temperature (25°C), are presented. Fig. 3B shows typical stick-slip curves for two different fibers A and E, obtained by plotting force as a function of the vertical displacement of the horizontal fiber. The stick-slip events for fiber E, are characterized by a greater average amplitude and a smaller frequency than those of fiber A. However, when each friction curve is considered in more detail, stick-slip irregularities along a filament as well as from one filament to another are observed. To obtain a 95% confidence level, friction measurements were carried out on four different segments of 3 mm long filaments from five randomly chosen filaments.

Static friction coefficient (μ_s)

Fig. 9 shows the histograms of frequency distribution of static friction coefficients "μ_s" of fibers A and E respectively. For fiber E, the "μ_s" values follow a normal distribution with a mean value around 5 and a standard deviation of 1.7. However, for fiber A, a bimodal distribution of "μ_s" values is observed, with a mean value of the first distribution situated around 2 and that of the second distribution around 6.

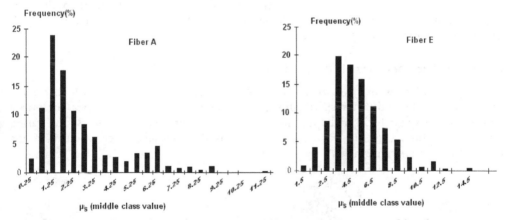

Fig. 9. Histogram of static friction coefficients of fibers A and E, measured by electronic microbalance

3.3 Comparison of fiber-fiber friction to friction analysis of surfaces by LFM

The greater static friction coefficient values of fiber E (than the fiber A), during fiber-fiber friction can be explained by the greater frictional values detected by the LFM technique. However, as the LFM signals reveal no chemical heterogeneity of the fiber E surface (constant friction coefficient of 0.2), the great standard deviation of friction values during fiber-fiber friction can be attributed to the surface roughness of fiber E: this leads to different contact area each time a fiber is in contact with an another one.

In the case of fiber A, in addition to surface roughness, chemical heterogeneity due to two regions with different friction coefficients (0.04 and 0.12) was observed by LFM. The great standard deviation of "μ_s" as well as its bimodal distribution, during fiber-fiber friction are thus due to surface roughness and chemical heterogeneity.

The low friction coefficient values obtained by LFM may be explained by the fact that sliding of AFM tip on a fiber sample surface is easier ($\mu < 1$), than that of a fiber on another fiber surface ($\mu_s \sim 1$-10). Moreover in LFM, the friction force analyzed is a dynamic one, while in this case, it is the static friction coefficient which is being evaluated. The friction coefficient values disparities can be related to scale difference of measurements and to the nature and surface of contact which are different in both cases. In AFM, a Si_3N_4 tip having a radius of curvature of about 30 nm, is in contact with the fiber surface while during fiber-fiber friction measurements sized fibers of diameter 11 µm are in contact.

3.4 Determining the theoretical contact area during the fiber-fiber friction

The theoretical contact area during friction, can be calculated by using the DMT and JKR theories of contact mechanics. But according to Pashley, as the glass fiber is rigid and the fiber radius as well as its surface energy small, the DMT theory should be applied (Pashley, 1984). For orthogonally placed fibers, if contact occurs at one point, the contact radius derived by the DMT is 0.118 µm . However, contact between the two fibers occurs on a length of a circular arc of $2\pi r$ ($\theta/360$) (where the wrap angle θ=2.3° and the fiber radius: r=5.5 . 10^{-6} m (see Fig. 3). Thus the area of contact 'A' between the two fibers is approximately an ellipse, and it is evaluated to be A=0,12 µm² (details of this result is published in another paper (N. Behary et al., 2000).

3.5 Types of contact possible during the fiber-fiber friction of the fiber A

The AFM/LFM measurements of the fiber A revealed two regions of distinct friction coefficient ($\mu \sim 0.04$ and 0.12 respectively), each of them having a surface area comprised between approximately 0.1 and 1µm², that is nearly the same contact area as that during fiber-fiber friction. Thus, the main types of contacts that can take place would have the configurations illustrated in Fig. 10.

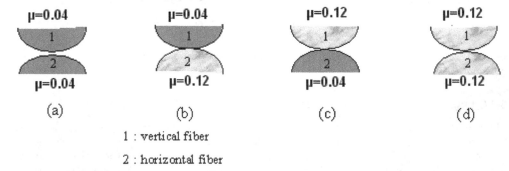

1 : vertical fiber

2 : horizontal fiber

Fig. 10. Modelling the different contacts possible during fiber-fiber friction of the sized fiber A (the friction coefficients are those evaluated by LFM)

Globally therefore, there can be four different types of contact during the fiber-fiber friction. We can compare these configurations to the results obtained by Yamaguchi (Yamaguchi, 1990) for polymer-polymer and polymer-steel friction force measurements. Yamaguchi observed that the friction coefficient of any polymer against PTFE (polytetrafluoroethylene) was nearly the same and of a small value (μ~0.1), while the friction of a polymer against steel varied a lot. In fact, it is the polymer presenting a weak friction coefficient that imposes the value of the relative friction coefficient (in the limit of the study considered). Therefore it can be concluded from our study that contacts a, b and c lead to a relative weaker friction coefficient than contacts of the type d. This would lead to a bimodal distribution of the relative friction coefficients during fiber-fiber friction and would explain our experimental results.

4. Conclusions

In light of the above results and discussions, both techniques of measuring friction forces, by Lateral force microscopy and an electronic microbalance, seem to be invaluable methods for characterizing frictional properties of sized glass fibers. The AFM/LFM successfully determines the topography and chemical nature of sized and desized glass fibers. Sized fibers have both physical and/or chemical heterogeneities while the desized bare glass fiber is completely plain and smooth.

AFM/LFM results also help to better understand friction force results obtained at a larger macro scale, particularly the widespread values of friction coefficients during fiber sliding. Nanoscale friction values by AFM/LFM are smaller than 'micro' friction values during fiber-fiber friction because the nature and the area of contact are different in both cases.

Part II: Air-atmospheric plasma treatment of PET (Polyethylene Terephtalate) woven fabrics studied by atomic force microscopy

1. Introduction

Polyester fabrics made from PET poly(ethylene terephtalate) account for almost 50% of all fiber materials. PET fibers have high uniformity, mechanical strength or resistance against chemicals or abrasion. However, high hydrophobicity, the build-up of static charge, stain retention during laundering and being difficult to finish are undesirable properties of PET.

Enhancement of the hydrophilicity of PET fibers is a key requirement for many applications, ranging from textile production to applications in the biomedical field. In the textile field, increased hydrophilic properties improves comfort in wear with a better moisture management due to increased wettabilility, wicking, adhesion to other materials (i.e. coating), and dyeing (Pastore M, 2001). Several strategies can be adopted to increase the surface energy and hence the hydrophilicity of PET fibers such as by chemical finishing or grafting, chemical surface treatment with NaOH (Collins, 1991, Haghighatkish, 1992), biochemical treatment with enzymes (Vertommen, 2005) or physical surface treatment using plasma. Treatment with NaOH is environmentally unfriendly and causes drastic weight and strength losses (Collins, 1991), while certain plasma treatments would be of interest from an environmental point of view.

Plasma techniques have been used in materials science since 1960s, for the activation and modification of different materials. Plasma processes have been utilized to improve the surface properties of fibers for various textile applications: sterilization, desizing, wettability or hydrophobicity improvement, anti-shrinking finishing, dyeability enhancement and adhesion promotion (Hocker 2002, Jasso, 2006, Leroux, 2006, Oktem, 2000). The fibers that can be modified by plasma processing include almost all kinds of fibers such as natural and man-made, metallic fibers, glass fibers, carbon fibers and organic fibers (Höcker, 2002, Borcia, 2003). Low-pressure plasma methods have been investigated, but they are difficult to apply in industry since they require vacuum and consume a considerable amount of energy. Moreover, these treatments can be carried out only in a batch process which increases the treatment time. New methods based on atmospheric plasma treatments seem to be quite attractive for the textile industry. These treatments have the advantage of being applied on-line without vacuum, and allow continuous plasma processing (Leroux, 2006). Atmospheric plasma treatments are used to modify polymer surfaces using plasma gases made up of a mixture of charged particles (electrons and ions), excited atoms (free radicals, meta-stable molecules) and photons. In order to create a plasma field, the gases are brought through two charged electrodes of different potentials. During plasma treatment, the polymer to be treated is exposed to the plasma which interacts and modifies the polymer surface. Surface modifications vary with the nature of the substrate and the chemistry of plasma gases, as well as the treatment operating parameters (Borcia, 2004, Leroux, 2006,)].

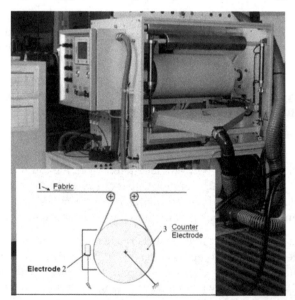

Fig. 11. Plasma treatment under atmospheric pressure by means of dielectric barrier discharge, using "Coating Star" plasma machine manufactured by the Ahlbrandt System Company

Among various atmospheric pressure non-thermal plasmas, the dielectric barrier discharge (DBD) process is studied mostly for the easy formation of stable plasmas and its scalability. Numerous studies of surface modified materials using DBDs under atmospheric pressure have been undertaken mainly for treatment of polymers and metals. The dielectric barrier discharge (DBD) is created between two metal electrodes covered with insulating layers (ceramic and polycarbonate). Under specific conditions, the DBD can produce the so-called atmospheric pressure glow (APG) discharge, which can also be effectively used for the surface treatment (Borcia, 2003, Fang, 2004).

Studies show that the highly reactive species in the discharge regime of DBDs can interact with the surfaces of materials and induce some physical and chemical changes (oxidation, polymerization, cross-linking, etching etc) and thus improve their surface properties such as wettability, printability, adhesion and conductivity (Oktem, 2000, Fang, 2004).

Ageing of surface-modified polymers can be detrimental to the performance of a device. Therefore an improved understanding of the ageing of modified surfaces during storage is required to optimize the processing conditions leading to interfacial properties that are controlled and predictable at the time of use. It has been shown that as a result of ageing, the surface properties, acquired after a plasma treatment, disappear (Krump, 2005). Moreover, ageing survey data (Borcia, 2004) show that materials never fully return to their untreated surface state. The ageing effects observed depend on the nature of the polymeric materials, structure, crystallinity, (Novak, 2004) porosity etc. Additionally, it has been reported (Leroux, 2006), that the fabric porosity which depends on the fabric structure (e.g. woven, nonwoven, knitted) influences the ageing of air-atmospheric plasma treated fabrics subjected to aqueous conditions at room temperature and pressure. It was also shown that for a low porosity-(high density) woven fabrics, oxidised species formed at the fabric surface are more easily removed by simple washing than for high porosity textile structure.

The work presented here demonstrates the potential use of Atomic Force Microscopy to optimize the atmospheric plasma process parameters (speed treatment and electrical power) on the treatment of a particular woven PET fabric. The effect of ageing during normal use conditions (i.e room temperature and pressure in absence and presence of day light) and in conditions used for dyeing polyester fabrics, has equally been studied. PET fabrics are dyed with disperse dye and need high temperature (130°C) and pressure for maximum dye uptake. Dyeing at 90°C would be more environmentally friendly because of reduced energy requirements, but dye uptake would be reduced. One of the strategies to increase dye uptake at 90°C is to activate the PET fabric surface by plasma treatment so as to increase its hydrophilicity. It is therefore necessary to check the permanence of the air-atmospheric plasma treatment of PET fabrics for specific dyeing conditions.

In the final part, the effect of immobilizing a hydrophilic oligomer PEG -poly(ethylene glycol) 1500 immediately after plasma treatment is shown. AFM results are also compared to other fabric surface characterization methods such as water contact angle, capillary uptake, and XPS (X-Ray Photoemission Spectroscopy).

2. Experimental work

2.1 Sample preparation

2.1.1 Woven polyester fabric

A 100 % Polyester (PET) woven fabric of density 284 g/m² with a thickness of 0.56 mm and 63.5% porosity was used for the study. The PET woven fabric was cleaned to be free from surface impurities and spinning oil. The cleanliness of the PET samples was checked by measuring the surface tension of final rinsing water (used to clean the PET samples), which remained constant and equal to 72.6 mN/m ,which is the surface tension of pure water.

2.1.2 Sample preparation for plasma treatment

The PET woven fabric was cut into square pieces of 50cm X 50cm on the basis of the electrode length of the plasma machine (50 cm). The speed of the fabric in the discharge zone could be varied through the control.

2.1.3 Plasma treatment of woven PET samples

All plasma treatments were carried out using an atmospheric plasma machine (see Fig. 12) called 'Coating Star' manufactured by Ahlbrandt System (Germany). The following machine parameters were kept constant: electrical power of 1 kW, frequency of 26 kHz, electrode length of 0.5 m and inter-electrode distance of 1.5 mm. The outer layer surfaces of both electrodes were of ceramic (a dielectric material), so that when these electrodes were subjected to a potential difference, a glow discharge called the 'Dielectric Barrier Discharge' (DBD) was created.

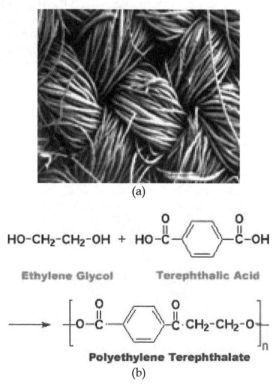

(a)

Ethylene Glycol   Terephthalic Acid

Polyethylene Terephthalate
(b)

Fig. 12. (a) Optical microscopic view of the PET woven fabric, (b) chemical formula of PET

Atmospheric air was used during the atmospheric plasma treatments. The textile samples were subjected to varying plasma Treatment Power (TP) which is the plasma power applied per m² of textile sample, expressed in kJ/m². The TP is related to the velocity of the treatment (V) and the electrical power (P) of the machine, by the equation (Eq. 1):

$$TP = \left(\frac{P}{V \times L} \right) \times 0.06 \qquad (1)$$

P = Electrical Power (W)
L = Electrode length (m)
V = Velocity of the sample (m/min)

Plasma treatment was carried out at velocities (V) of 1 m/min, 2 m/min, 5 m/min and 10m/min. It was also performed at constant speed with varied electrical power (P) of 400 watts, 700 watts and 1000 watts. After plasma treatment, each plasma treated sample was separated from waste fabric and kept in aluminium foil away from light.

2.1.4 Ageing methods

Effect of ageing on air-plasma treated PET fabric samples was observed by monitoring changes in water contact angle as well as surface topography of the samples, 20 days after plasma treatment. Ageing of plasma treated sample was carried out in two different ways:

- *Without light:* Plasma-treated samples were kept folded in aluminium foil and stored in cupboard, in a small dark chamber to avoid contact with light.
- *With light:* Plasma treated samples were kept open in laboratory conditions allowing them to be in contact with light. Laboratory conditions with a temperature of 20 ± 2°C and relative humidity of 60± 10 % were maintained during the experiment.

2.1.5 Effect of PET fabric dyeing conditions (High temperature conditions)

The plasma treated PET samples were immersed in aqueous conditions at 90°C for 30 minutes without the presence of any dye.

2.1.6 Effect of adhesion of PEG 1500 on PET fabric

As ageing with time causes loss of hydrophilic species formed by plasma treatment (Krump, 2005), immobilizing a hydrophilic oligomer like PEG -poly(ethylene glycol) immediately after plasma treatment would perhaps yield a more durable hydrophilic treatment. PEG-poly(ethylene glycol) has been used for surface modification because of its unique properties such as hydrophilicity and flexibility (Harris, 1992).

PEG of molecular weight 1500 g/mol , i.e PEG 1500 from Fluka chemicals was immobilized on cleaned-untreated PET fabrics as well as plasma treated PET fabric samples using padding and curing method. For the padding process the open-width PET fabric was passed through an aqueous solution of PEG 1500 in water and through two squeezing rollers (see Fig. 13). At a squeezing pressure of 4 bars, the weight pick up remained almost constant around 56%.

2.2 AFM tapping mode images

In a previous paper Leroux and al. (Leroux, 2006) showed using AFM in the LFM mode, that friction forces measured at the surface of PET fabric is doubled after a plasma treatment, and these forces remain homogeneous throughout the plasma treated PET fiber surface.

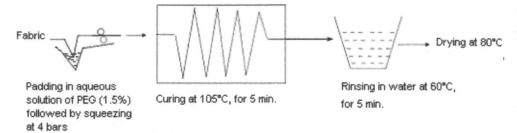

Fabric

Padding in aqueous
solution of PEG (1.5%)   Curing at 105°C, for 5 min.
followed by squeezing
at 4 bars

Rinsing in water at 60°C,
for 5 min.

Drying at 80°C

Fig. 13. Application procedure of PEG-1500 on PET fabric by padding, squeezing followed by curing and a final rinsing before drying.

In this chapter, surface investigation by imaging in the Tapping mode using AFM "Nanoscope III" from Digital Instrument, is presented. This mode was preferred to the AFM /LFM (contact mode AFM), since the Tapping mode overcomes problems associated with friction, adhesion, electrostatic forces which may arise after a plasma treatment, and which would distort image data (Kailash, 2008).

Tapping mode imaging was carried in ambient air using 'Budget sensor' tips from 'Nanoandmore', of length 125 µm, made of $Si_3 N_4$ with aluminium coating, and a resonance frequency of 300 kHz. The cantilever which scans the surface is oscillated at or near its resonant frequency using a piezoelectric crystal. The oscillating tip is moved toward the surface until it begins to lightly touch, or tap the surface, significantly reducing the contact time. During scanning, the vertically oscillating tip alternately contacts the surface and lifts off, generally at a frequency of 5000 to 500,000 cycles per second. As the oscillating cantilever begins to intermittently contact the surface, the cantilever oscillation is necessarily reduced due to energy loss caused by the tip contacting the surface. The reduction in oscillation amplitude is used to identify and measure surface features. When the tip passes over a bump in the surface, the cantilever has less room to oscillate and the amplitude of oscillation decreases. Conversely, when the tip passes over a depression, the cantilever has more room to oscillate and the amplitude increases (approaching the maximum free air amplitude). This feed back signal also allows construction of the topographic image. The fiber surface roughness Ra was calculated directly from AFM signals (Feninat, 2001) using the software supplied with the Nanoscope III.

2.3 Water contact angle and capillary measurements

The sessile drop method is useful for measuring water contact angles greater than 90° however for lower contact angles the porous woven fabric immediately absorbs the liquid drop. That is why wettability of textile surfaces was monitored indirectly through a wicking test using a tensiometer. This method allows the measurement of contact angle as well capillarity measurements. The tensiometer used is the 3S from GBX, France. During measurements, a vertically hanging fabric sample of defined dimension is connected to the tensiometer at the weighing position and progressively brought into contact with the surface of water placed in a container. On immediate contact with the water surface, a sudden increase in weight (W_m) is measured due to meniscus formation on the fabric surface. The weight increases further as the liquid flows inside the fabric structure by capillarity (wicking). The capillary weight (W_c) after 6 minutes, is measured. From the meniscus weight (W_m) the water contact angle can be calculated. More detailed description of this experiment can be found in our previous work (Leroux, 2006).

3. Results and discussions

3.1 Effect of varying speed and electrical power on plasma-treatment of PET woven fabrics

Irrespective of treatment power and speed setting, the water contact angle after plasma treatment was always around 45°, while the untreated PET fabric water contact angle was 80°. Any decrease in speed or increase in Treatment Power did not have a significant impact on this minimum contact angle value. However, tapping mode AFM images of samples (Fig. 14) subjected to an air–plasma treatment at varying Treatment powers "TP" of 0, 12, 24, 60 and 120 kJ/m² show considerable changes in fiber surface morphology.

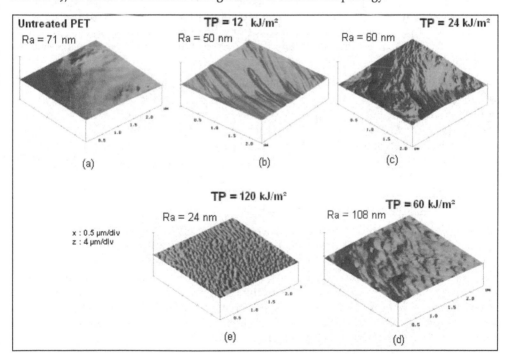

Fig. 14. Topographic images obtained by AFM in the tapping mode of a) an untreated PET fabric fiber surface, a plasma-treated PET fabric fiber surface at treatment power "TP" of (b) 12 kJ/m², (c) 24 kJ/m² , (d) 60 kJ/m² and (e) 120 kJ/m² (Takke, 2009).

Fig. 14 shows that as the treatment power is increased from 0 to 120 kJ/m² there is significant surface restructuring. At very low Treatment Power of 12 kJ/m² at a speed of 10 m/min, a smoother surface is obtained observable as a decrease in surface roughness, Ra=50 nm, (Fig. 14 b). The smoother surface at this highest speed and low plasma power, may be due to etching of the very upper layer of the PET surface. However, as the Treatment Power is further increased to 24 KJ/m^2 at a lower speed of 5 m/min, further etching of the PET surface creates disordered bumps (Fig. 14c). Continued increase of TP to 60 KJ/m^2 at a lower speed of 2 m/min, leads to more uniform scale-like structures (6 scales per μm, Fig. 14d) which increases the surface roughness of the fiber, Ra= 108 nm. Further increase in TP to 120 KJ/m^2 (at 1 m/min) causes further uniform etching yielding tinier, more organized, uniform and flatter beads (12 beads/μm, Fig. 14e) leading to a decrease in the surface roughness (Ra=24 nm).

The increase in hydrophilicity of the PET fabric surface after the atmospheric-air plasma treatment is probably due to plasma oxidation, which destroys the surface chemical bonds, leading to an increase in polar groups (Borcia, 2003). Plasma treatments generate polymer chain-scissions of the weakest bonds of the polyester, creating very reactive chain-end free radicals. These radicals then react easily with the reactive species (ex: oxygen radicals) present in the plasma generating polar species such as carbonyl, carboxyl and hydroxyl groups, which are polar species capable of increasing the surface energy (Leroux, 2006). We have previously reported using X.P.S measurements (X-Ray Photoemission Spectroscopy) (Leroux, 2009) that there is an increase in concentration of oxygen at the fabric surface after air plasma treatment of the PET fabric, and this would explain the increase in surface energy of the fabric surface.

The plasma treatment not only causes chemical modifications of the PET surface by adding polar groups, but also morphological surface changes that are observed by tapping mode AFM imaging. Although the lower water contact angle (45°) is measured even at low   Treatment Power (12 kJ/m^2), surface etching of the PET fiber surface continues with increasing Treatment Power, yielding an increased surface roughness. At higher Treatment Powers (60 and 120 kJ/m^2) a more organized scale–like surface structure is observed

3.2 Effect of ageing on plasma treated PET sample

Ageing of atmospheric plasma treated PET woven fabric with TP of 60 KJ/m^2 in the absence and presence of light was evaluated. In absence of light, very little change in water contact angle was observed. However, there were substantial increase in water contact angle (45° to 73°) and decrease in capillary weight for plasma treated PET fabrics kept in presence of daylight. The increase in contact angle with time, in presence of light, could be due to the loss of surface oxidation species at the plasma treated PET fabric surface. Degradation of peroxide radicals due to UV rays of the daylight could be a major factor in increasing the water contact angle of the plasma treated fabric (Takke, 2009).

Tapping mode AFM imaging carried out after the 20th day of ageing in the absence of light shows that there is deformation of the scale-like structures formed as a result of plasma treatment at 60 KJ/m^2 observed as a disordered surface structure with big bumps. In presence of light, smaller, flatter and fewer blister-shaped structures, of different sizes, appear as if, the scales formed by plasma treatment had been nearly completely eroded, leaving a smoother (Ra =52 nm).

While in absence of light slow ageing initiates a disordering of the surface, ageing in presence of light leads to an organized structure. Without light, the ageing process proceeds slowly as evidenced by the similar water contact angles of the 20 and 1 day samples. As only slight change in the surface topography is observed, the oxidised species should stay on the fiber surface and are not removed by ageing without light. With light, the ageing process seems to be accelerated, both at the outer fabric surface and the inner fabric fiber surface.

Further chemical analysis of the surfaces for example by XPS (X-Ray Photoemission Spectroscopy) would provide clearer information on the effect of light on plasma treated samples, however AFM imaging seem to be an invaluable tool to investigate ageing effect on surface structuring.

3.3 Effect of PET fabric dyeing conditions (High temperature conditions)

Plasma treatment alone increases the hydrophilicity of PET fabric, demonstrated by the decrease in water contact angle from 80° to 45°. However, after 30 minutes of immersion in water at R.T.P there is an increase in water contact angle from 42° to 55°, and a considerable decrease in capillary weight from 360 to 80 mg. Thus, the polar species formed as a result of plasma treatment are gradually lost during washing even at R.T.P. Under dyeing conditions, that is washing at 90°C for 30 min, the wettability of the plasma treated PET reaches that of untreated PET fabric (water contact angle=80°) meaning that there is complete loss of all polar species formed by plasma treatment. Surface analysis of plasma treated PET subjected to hot aqueous dyeing conditions shows complete disappearance of the scaly shaped PET surface (Fig. 15b). Only traces or finger prints of the scales remain leaving a smooth surface (Ra = 38 nm). Moreover dyeing of plasma treated PET shows no increase in disperse dye uptake.

3.4 Effect on adhesion of PEG 1500

Atomic Force Microscopy, in the Tapping mode was carried out to better understand the different morphological changes occurring at the PET fabric fiber surface before and after PEG-1500 immobilization with and without a prior plasma treatment of the PET fabric surface. Detailed results have been published elsewhere (Takke, 2010).

Figures 16a and 16b show typical topographical images of a cleaned-untreated PET fiber with and without PEG coating respectively. Though the surface roughness of the PET surface with or without PEG coating is almost the same (~70nm), the untreated PET fiber seem quite smooth and homogeneous. However, after application of PEG, many small thin elliptical and irregular shaped deposits, most probably due to PEG, appear at the fiber surface (see Fig. 16b).

After plasma treatment of PET fiber (at 60 kJ/m²) uniform scale-shaped bumps (6 scales per µm in the x-direction) appear on the PET fiber surface (see Fig. 16c) as a result of surface etching caused by plasma treatment. After immobilizing of PEG 1500, big bumps appear in the foreground relief, covering the regular scaly shaped bumps which can no longer be perceived. However, on the upper part of the image (x =0 to 1), the scaly shaped bumps still appear as if the gaps (valleys) in between scales have not been completely filled with PEG in that region, most probably, because either the PEG does not cover them at all, or only a thin layer of PEG covers the scaly bumps.

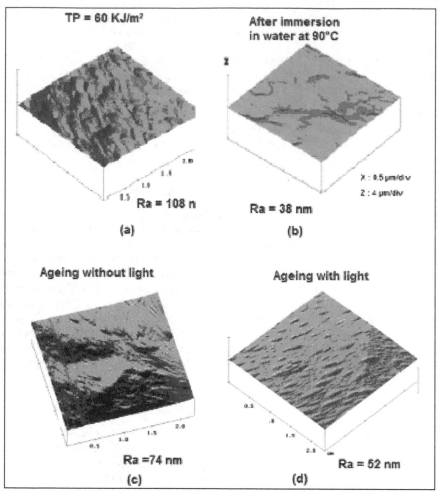

Fig. 15. Tapping mode topographic image of (a) a PET plasma-treated sample at 60 kJ/m², (b) followed by immersion in hot water at 90°C during 30 minutes, (c) after having been subjected to a 20 day-ageing without light and (d) ageing with light

Immobilisation of PEG on both untreated –cleaned PET and plasma treated PET lead to a hydrophilic fabric surface with similar water contact angle (~50°) in both cases. However, wash fastness test carried out at room temperature on PEG coated PET fabrics shows that without plasma treatment there is an increase in water contact angle of the PET fabric with an increase in washing cycles: the hydrophilic PEG molecules adsorbed at the PET surface are gradually desorbed with the washing time. This would mean that only weak cohesive forces exist between the cleaned untreated PET surface and the PEG molecules.

However, a plasma treatment of the PET fabric prior to PEG adsorption improves adhesion between PEG and PET-polyester fabric, since the water contact angle value of the PEG coated fabric remains unchanged with washing time at R.T.P and at 80°C. Therefore, the

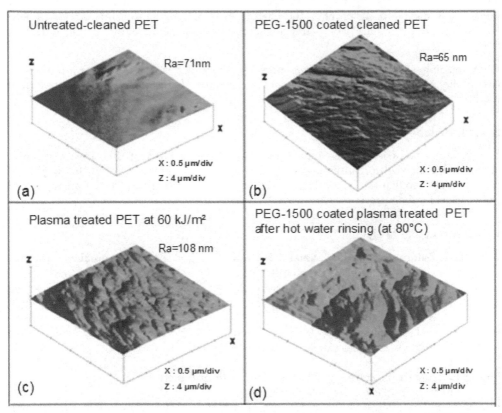

Fig. 16. Tapping Mode AFM Topographical images of (a) an untreated-cleaned PET fiber, (b) a PEG-1500 coated untreated-cleaned PET fiber, (c) a plasma treated PET fiber at 60 KJ/m², (d) a PEG-1500 coated plasma treated PET fiber subjected to wash fastness test at 90°C (Takke, 2011).

PEG molecules are still attached to the plasma treated PET fiber surface and cannot be removed by washing. The chemical or physico-chemical interaction/bonding between the free PEG and the polar species of the plasma treated PET results in an increased adhesion between PEG and PET fabric surface. It is also probable that the increase surface roughness (Ra =108 nm, as shown by AFM images) would enhance this adhesion.

4. Conclusions

AFM seems an invaluable tool to detect all surface morphological modifications taking place during plasma treatment of a PET fabric surface. It confirms that plasma treatment not only adds polarity to the PET surface, but also, depending on the treatment power used, etching of the PET surface by plasma or a reorganisation of the PET surface takes place. AFM also shows the changes that takes place after ageing in presence or in absence of light, and when the plasma treated fabric is subjected to high temperature aqueous conditions. It confirms that the loss of polarity during ageing is also accompanied by morphological changes. AFM imaging can also be used to obtain the fiber surface topography after the fabric has been padded with a hydrophilic PEG coating and shows the distribution of the coating at the PET fiber surface.

AFM imaging seems to be complementary to the other surface characterisation tools (wettability, XPS) and can be successfully used to characterise textile fiber surface modifications by plasma treatment

5. Acknowledgement

The authors acknowledge Asia Link Project for the realization of this project. We also acknowledge Utexbel for providing us with woven fabric sample used in this work. We would also like to thank Mr. Dassonville, Mr. V. Takke, Dr. F. Leroux, Dr. C. Campagne and Mr. C. Catel for their precious help.

6. References (Part I)

Baselt D.R., Baldeshwielder J.D., "Lateral Force Microscopy of Graphite in air", *J. Vac. Sci. Tech.* B-10 (5), 2316-2322 (Sept/Oct 1992).

Behary, N. et al., "Tribological analysis of glass fibers using atomic force microscopy (AFM)/lateral force microscopy (LFM)", Journal of Applied Polymer Science,75(8), 1013–1025 (2000)

Behary, N. et al., Tribology of Sized Glass Fibers, Part I: Friction Analysis by Lateral Force Microscopy and Electronic Microbalance Technique, *Text. Res. J.*, 78 (8), 700-708 (2000)

Bining, G., Quate, C.F., & Gerber, Ch., "Single-tube three-dimensionnal scanner for scanning microscopy", *Phys. Rev. Lett.* 56, 930 (1986).

Derjaguin, B.V., Muller, V.M., Toporov Yu.P., "Effect of contact deformation on the adhesion of particles" *Colloid Interface Sci.*, 53, 314-326 (1975)

Gupta, B.S., "*Frictional Behaviour of Fibrous Materials* ", *Polymer and Fiber Science, Recent Advances*, Edited by Raymond E., Fornes and Richard D. Gilbert, 305-332 (1992)

Gupta, B.S., Wolf, K.W., "Effect of Suture Material and construction on Frictional properties of sutures", *Gynecol. Obster.*, 161, 12-16 (1985)

Gupta, S., Molgahzy, E., "Friction in fibrous materials", Part II : Experimental study of effects of structural and Morphological factors", *Text. Res. J.*, 63 (4), 219-230 (1993)

Israelachvili, J., Chen, Y.L, "Fundamental Mechanisms of Interfacial Friction, 1.. Relation between Adhesion and Friction", *J. Phys. Chem.*, 97, 4128-4140 (1993).

Johnson, K.L., Kendall, K.,& Roberts, A.D, "Surface energy and contact of elastic solids", *Proc. Roy. Soc. London*, A324, 301-313 (1971)

Marti, O., " Nanotribology : Friction on a Nanometer Scale", *Physica Scripta* T 49, 599-604 (1993).

Meyer, E., and Amer, N.M., "Simultaneous measurement of lateral and normal forces with optical-beam-deflection atomic microscope", *Appl. Phys. Lett.*, 57(20), 2089-2090 (1990).

Pashley, M.D., Tabor, D., "Adhesion and micromechanical properties of metal surfaces", *Wear* 100, 7 (1984)

Plueddeman, P., "*Silane Coupling agents*", *Plenum Press, New York* (1982)

R. Overney, and E. Meyer, , "Tribological Investigations using Friction Force Microscopy," *Mrs Bulletin*, 26-34 (May 1993).

Williams J.A., "Engineering tribology" , *Oxford science publications* (1994)

Yamaguchi Y., "Tribology of plastic materials", *Tribology of plastic materials,* Elsevier, Amsterdam (1990)

7. References (Part II)

Borcia G., Anderson C. A. and Brown N. M. D., The surface oxidation of selected polymers using an atmospheric pressure air dielectric barrier discharge. Part I. *Appl. Surface Sci.* 2004, 221, 203

Borcia G., Anderson C. A. and Brown N. M. D., The surface oxidation of selected polymers using an atmospheric pressure air dielectric barrier discharge. Part II, *Appl. Surface Sci.* 2004, 225, 186

Borcia G., Anderson C. A., and. Brown N. M. D, *Plasma Sources Sci. T."* Dielectric barrier discharge for surface treatment: application to selected polymers in film and fiber form . 2003;12:335-344 T

Chatelier, C., Xie, X., Thomas, R., Gengenbach, and Griesser, Hans J., "Effects of plasma modification conditions on surface restructuring", *Langmuir* 1995;11:2585-2591.

Collins M.J., Zeronian S.H., Semmelmeyer M., "The use of aqueous alkaline hydrolysis to reveal the fine structure of poly(ethylene terephthalate) fibers", *J. Appl. Polym. Sci.,* 1991, 42, 2149

Fang, Z., Qiu, Y., and Kuffel, E., Formation of hydrophobic coating on glass surface using atmospheric pressure non-thermal plasma in ambient air. *J. Phys. D: Appl. Phys.* 2004; 37:2261-2266.

Haghighatkish M., Youscfi M., "Alkaline hydrolysis of polyester fibers. Structural effects", *Iranian Journal of Polymer Science & Technology",* 1992, 1(2), 56

Harris JM, editor. Poly(ethylene glycol) chemistry: Biotechnical and Biomedical applications (1992), New York: Plenum Press.

Höcker H., Plasma treatment of textile fibers, *Pure Appl. Chem.* 2002, 74(3), 423–427.

Jasso, M., Hudec, I., Alexy, P., Kovacik, D., Krump H., "Grafting of maleic acid on the polyester fibers intiated by plasma atmospheric pressure" *Int. J. adhes.* Adhes. 2006; 26:274-284.

Kailash C. Khulbe,C. Y. Feng,Takeshi Matsuura, Synthetic Polymeric Membranes: Characterization by Atomic Force Microscopy, GERMANY, Springer-Verlag Berlin Heidelberg (2008)

Krump, H., Simor, M., Hudec I., Jasso, M., Luyt A. S., Adhesion strength study between plasma treated polyester fibers and a rubber matrix, *Appl. Surf. Sci.*2005; 240:268-274.

Leroux, F., Perwuelz, A., Campagne C., Behary N., "Atmospheric air plasma treatments of polyester textile structures ", *J. Adhesion Sci. Technol.* 2006;20(9):939-957.

Leroux F., Campagne C., Perwuelz A., Gengembre L., "Fluorocarbon nano-coating of polyester fabrics by atmospheric air plasma with aerosol", *Appl. Surf. Science,* 2008, 254(13), 3902-3908.

Novak, I., Florian S., "Investigation of long-term hydrophobic recovery of plasma modified polypropylene ", *J. Material Sci.* 2004; 39(6) :2033-2036.

Oktem, T., Seventekin, N., Ayhan, H., Piskin E., "Modification of polyester and polyamide fabrics by different in situ plasma polymerization methods", *Turk J Chem*. 2000; 24 : 275-285.Pastore M. and Kiekens P. (2001) In: Surface Characteristics of fibers and textiles, edited by Christopher Pastore, Surfactant Science Series, Vol 9, 298 pages

Takke V., Behary N., Perwuelz A., Campagne C., "Studies on the atmospheric air–plasma treatment of PET (polyethylene terephtalate) woven fabrics": Effect of process parameters and of aging, *J. Appl. Polym. Sci.* 2009, 114(1), 348

Takke V., Behary N., Perwuelz A., Campagne C., Surface and adhesion properties of poly(ethylene glycol) on polyester(polyethylene terephthalate) fabric surface: Effect of air-atmospheric plasma treatment, *J. Appl. Polym. Sci.* 2011, 122 (4), 2621

Verschuren J., Van Herzele P., Clerck, K. De., and Keikens, P, Influence of Fiber Surface Purity on Wicking Properties of Needle-Punched Nonwoven after Oxygen Plasma Treatment, *Text. Res. J.*, 2005; 75(5):437-441.

Vertommen M.A.M.E.,. Nierstrasz V.A, Van der Veer M.,Warmoeskerken M.M.C.G., "Enzymatic surface modification of poly(ethylene terephthalate)", *J. Biotechnol.* 2005, 120, 376

Permissions

The contributors of this book come from diverse backgrounds, making this book a truly international effort. This book will bring forth new frontiers with its revolutionizing research information and detailed analysis of the nascent developments around the world.

We would like to thank Victor Bellitto, for lending his expertise to make the book truly unique. He has played a crucial role in the development of this book. Without his invaluable contribution this book wouldn't have been possible. He has made vital efforts to compile up to date information on the varied aspects of this subject to make this book a valuable addition to the collection of many professionals and students.

This book was conceptualized with the vision of imparting up-to-date information and advanced data in this field. To ensure the same, a matchless editorial board was set up. Every individual on the board went through rigorous rounds of assessment to prove their worth. After which they invested a large part of their time researching and compiling the most relevant data for our readers. Conferences and sessions were held from time to time between the editorial board and the contributing authors to present the data in the most comprehensible form. The editorial team has worked tirelessly to provide valuable and valid information to help people across the globe.

Every chapter published in this book has been scrutinized by our experts. Their significance has been extensively debated. The topics covered herein carry significant findings which will fuel the growth of the discipline. They may even be implemented as practical applications or may be referred to as a beginning point for another development. Chapters in this book were first published by InTech; hereby published with permission under the Creative Commons Attribution License or equivalent.

The editorial board has been involved in producing this book since its inception. They have spent rigorous hours researching and exploring the diverse topics which have resulted in the successful publishing of this book. They have passed on their knowledge of decades through this book. To expedite this challenging task, the publisher supported the team at every step. A small team of assistant editors was also appointed to further simplify the editing procedure and attain best results for the readers.

Our editorial team has been hand-picked from every corner of the world. Their multi-ethnicity adds dynamic inputs to the discussions which result in innovative outcomes. These outcomes are then further discussed with the researchers and contributors who give their valuable feedback and opinion regarding the same. The feedback is then collaborated with the researches and they are edited in a comprehensive manner to aid the understanding of the subject.

Apart from the editorial board, the designing team has also invested a significant amount of their time in understanding the subject and creating the most relevant covers. They scrutinized every image to scout for the most suitable representation of the subject and create an appropriate cover for the book.

The publishing team has been involved in this book since its early stages. They were actively engaged in every process, be it collecting the data, connecting with the contributors or procuring relevant information. The team has been an ardent support to the editorial, designing and production team. Their endless efforts to recruit the best for this project, has resulted in the accomplishment of this book. They are a veteran in the field of academics and their pool of knowledge is as vast as their experience in printing. Their expertise and guidance has proved useful at every step. Their uncompromising quality standards have made this book an exceptional effort. Their encouragement from time to time has been an inspiration for everyone.

The publisher and the editorial board hope that this book will prove to be a valuable piece of knowledge for researchers, students, practitioners and scholars across the globe.

List of Contributors

Martin Veis
Institute of Physics, Faculty of Mathematics and Physics, Charles University, Institute of Biophysics and Informatics, 1st Faculty of Medicine, Charles University, Czech Republic

Roman Antos
Institute of Physics, Faculty of Mathematics and Physics, Charles University, Czech Republic

F.A. Ferri, M.A. Pereira-da-Silva and E. Marega Jr.
Instituto de Física de São Carlos, Universidade de São Paulo, São Carlos, Brazil

M.A. Pereira-da-Silva
Centro Universitário Central Paulista, UNICEP, São Carlos, Brazil

Vishal Gupta
FLSmidth Salt Lake City Inc., USA

Thin-Lin Horng
Department of Mechanical Engineering, Kun-Shan University, Tainan, Taiwan, R.O.C.

Giovanna Malegori and Gabriele Ferrini
Interdisciplinary Laboratories for Advanced Materials Physics (i-LAMP) and Dipartimento di Matematica e Fisica, Università Cattolica del Sacro Cuore, I-25121 Brescia, Italy

R.R.L. De Oliveira, D.A.C. Albuquerque, T.G.S. Cruz, F.M. Yamaji and F.L. Leite
Federal University of São Carlos, Campus Sorocaba, Brazil

Marius Enachescu
University POLITEHNICA of Bucharest, Romania

Victor J. Bellitto
Naval Surface Warfare Center, USA

Mikhail I. Melnik
School of Engineering Technology, Southern Polytechnic State University, Marietta, GA, USA

Z. Chen
Golden Sand River California Corporation, Palo Alto, California, USA

L.W. Su, J.Y. Shi, X.L. Wang, C.L. Tang and P. Gao
Lattice Power Corporation, Jiangxi, China

Prabir Kumar Das, Denis Jelagin, Björn Birgisson and Niki Kringos
Railway and Highway Engineering, KTH Royal Institute of Technology, Sweden

Nemeshwaree Behary and Anne Perwuelz
ENSAIT-GEMTEX, ENSAIT, GEMTEX, Roubaix Univ Lille Nord de France, USTL, F-59655, Villeneuve d'Ascq Cedex, France

Printed in the USA
CPSIA information can be obtained
at www.ICGtesting.com
JSHW011444221024
72173JS00004B/935

9 781632 384096